POWER DEFINITIONS AND THE PHYSICAL MECHANISM OF POWER FLOW

POWER DEFINITIONS AND THE PHYSICAL MECHANISM OF POWER FLOW

Alexander Eigeles Emanuel
Worcester Polytechnic Institute, USA

IEEE
IEEE PRESS

WILEY
A John Wiley and Sons, Ltd., Publication

Registered office
John Wiley & Sons Ltd, The Atrium, Southern Gate, Chichester, West Sussex, PO19 8SQ, United Kingdom

For details of our global editorial offices, for customer services and for information about how to apply for permission to reuse the copyright material in this book please see our website at www.wiley.com.

Library of Congress Cataloging-in-Publication Data

Emanuel, Alexander Eigeles.
  Power definitions and the physical mechanism of power flow / Alexander Eigeles Emanuel.
    p. cm.
  Includes index.
  ISBN 978-0-470-66074-4 (cloth)
1. Electric charge and distribution–Mathematics. 2. Wave mechanics. 3. Electric power transmission. I. Title.
  QC581.E4E43 2010
  621.319–dc22   2010013501

A catalogue record for this book is available from the British Library.

ISBN: 978-0-470-66074-4

Typeset in 10/12pt Times by Aptara Inc., New Delhi, India
Printed in Singapore by Markono Print Media Pte Ltd

This book is dedicated
            to Rodica, my life-long love,
        and
            to three great teachers of mine:
            to my father Ing. Sigmund Eigeles
                    who taught me how to row a boat,
            to Professor Constantin I. Budeanu
                    who opened my eyes to the beauty
                            of energy conversion,
        and to Professor Michael S. Erlicki
                    who guided my first steps
                            in the realm of research.

Contents

Foreword

Energy policy is today a national priority. Conservation, renewable electricity sources and converting transportation from fossil fuels to electricity are major foci of activity. This is the most opportune time for someone to provide the definitive exposition of what is meant by "power" and how its various constituents are calculated and measured. Professor Alex Emanuel has done just that. This book, by gathering and clearly presenting the evolution of the increasingly complex interpretation of the V-I product, has done a great service to the engineering profession. Over time, the increased sophistication of electricity use and the introduction of solid state converters with their attendant harmonics have required that the definition and measurement of "power" be subjected to ever greater scrutiny. The resulting proposals, counterproposals and controversies evolved into a series of IEEE and DIN standards, each responding to the nascent effects of new technologies or an increased emphasis on economic accuracy. In today's restructured electricity industry, with real-time markets, the prospect of price responsive demand, and the introduction of an advanced metering infrastructure, accuracy of measurement and interpretation of power data is crucial.

Professor Emanuel promises an "in-depth understanding of the very physical mechanism that governs energy flow," and he delivers. He starts with the fundamental field description of power represented by Poynting's vector, then carefully and logically transitions to lumped physical systems, adding complexity until we arrive at all the components of "power" that constitute the V-I products of three phase, four-wire, unbalanced systems with non-sinusoidal currents and voltages. During the entire journey he builds upon the contributions of Budeanu, Fryze, Depenbrock and Czarnecki, pioneers in attempting to provide a rational interpretation of the result of multiplying current by voltage. And the comparisons he draws among the results of their proposed methods and that of the recent IEEE 1459-2010 standard is most interesting.

Few engineers are as well qualified as Professor Emanuel to undertake the writing of this monumental work. He has dedicated his entire professional life to the study of energy related problems, and has been a leading authority and educator in the field of power systems. It is clear that writing this book has been a labor of love, and we fellow electrical engineers owe Professor Emanuel a debt of gratitude for the fruit of this labor.

John G. Kassakian
Professor of Electrical Engineering and Computer Science
Laboratory for Electromagnetic and Electronic Systems
The Massachusetts Institute of Technology, USA

Preface

These days the implementation of an "Advanced Metering Infrastructure" will be determined by the technology that enables the manufacturing of "Smart Meters." The quest for the smart meter is an integral part of a revolutionary movement, the big impetus for the Smart Grid and energy management. When the flow and the use of electric energy is monitored by means of smart meters, the energy provider receives, via real-time data acquisition, information that enables the remote control of customer loads, the ability to adjust demand response, asset management, variable pricing programs, and many more capabilities that improve the energy transmission efficiency by reducing the power losses and by leveling the load curves. Smart meters also benefit customers by increasing the reliability of supply as well as the ability to monitor the use and cost of energy. This technology will stimulate much needed behavioral changes that will lead to a significant decrease in energy use.

The metering instrumentation is designed to conform to the mathematical definition of the electric quantity monitored; however, the definition on which such smart meters' design is based must be true to the law of physics and provide information that enables the accurate determination of energy flow rate and quality, optimum power dispatching, and efficient maintenance planning.

The meters in use today, even some of the most modern electronic meters, are designed and built following a tradition rooted in the 1930s and 1940s. It is well known that meters that measure energy (kWh) and active power (kW) provide accurate measurements also under nonsinusoidal or unbalanced conditions; nevertheless, meters dedicated to apparent power (kVA) and nonactive power (kvar) measurements are prone to significant errors when the current and voltage waveforms are distorted. The main reason for such uncertainties stems from the inadequate power definitions that dictate the conceptual design of such instrumentation. Evidently this situation led to the search for a practical solution. The progress toward universally accepted definitions is slow and hindered by economic factors tied to an existing infrastructure of large proportions. A lively debate over the apparent power definition and its resolution started a century ago and has not yet reached a conclusion. I am an active participant in this ongoing debate and have witnessed how, in the last decades, a vigorous "technological soul searching" has produced two significant standards:

1. The IEEE Std. 1459–2010, Definitions for the Measurement of Electric Power Quantities Under Sinusoidal, Nonsinusoidal, Balanced, or Unbalanced Conditions.
2. The DIN 40110-2: 2002–11, Quantities Used in Alternating Current Theory–Part 2: Multiconductor circuits (Polyphase Circuits).

Both standards provide improved power definitions that have been scrutinized and approved by large groups of experts and have triggered a multitude of engineering papers. This book has its main origin with the IEEE Std. 1459–2010. Many users of this document complained about its "tough reading" and asked for support documentation. Another important motivation for the production of this book stems from my desire to make a small contribution toward the acceptance and proliferation of smart meters. The major goals of this book are as follows:

1. To provide a clear understanding of the physical mechanism that governs the electric energy flow under different conditions: single- and three-phase, sinusoidal and nonsinusoidal, balanced, and unbalanced systems,
2. To be able to propose and advocate for recently developed power definitions that are not mathematical artifacts, but expressions that help correctly describe the actual effects and interactions between the energy sources, loads, equipment, and environment. Such definitions must be based on the solid understanding of the physical characteristics of the different components of energy,
3. To explain, discuss, and recommend power definitions that played a significant historical role in paving the road for the two standards,
4. To compare the two standards.

This book consists of eight chapters. The first explains electric energy flow. It introduces the concept of power as the rate of flow of energy and emphasizes the fact that electric energy is carried by an electromagnetic wave characterized by the power density (W/m^2) that is a function of time and space. Such electromagnetic waves travel (slide) along the conductors and contain a host of components. The main tool, used through the entire book, for recognizing the characteristics that help separate the actual elementary components of energy, is the Poynting vector. It is shown, with the help of a set of solved problems, that some components are active and carry unidirectional energy from sources to the loads, and other components oscillate between the loads and the sources and do not contribute to the net transfer of energy, but cause additional power loss in the conductors that connect loads and sources. This key chapter concludes that the Poynting vector is an excellent tool for the visualization of the power flow distribution in space and time; most importantly, it helps to reveal the correct energy and power components that ultimately are reflected in the ways the apparent power is resolved, thus leading to power definitions that are true to Nature's laws.

The second chapter deals with the single-phase system with sinusoidal waveforms. While introducing new definitions, such as the intrinsic power, this chapter presents a needed introduction and review of basic power definitions. Ample space is dedicated to the notion of apparent power, proving that it is a defined (convention) quantity that governs the equipment size, equipment losses, and the equipment aging and life-span. The power factor concept is presented in detail. A major section in this chapter is occupied by the discussions about the power oscillations between load and source, about the quantification of reactive and nonactive powers. It is shown that reactive power does not have to be produced by inductive or capacitive loads, but can also be caused by any energy converter that has the ability to store and return energy. It is also proved that the Poynting vector provides an excellent vehicle to help the reader familiarize with the separation and grouping of different active and nonactive instantaneous power components, their mathematical symbolism, and physical interpretation.

Chapter 3 explains the single-phase systems with nonsinusoidal waveforms. It starts by analyzing linear loads exposed to nonsinusoidal excitation and gradually advances to basic nonlinear loads, ultimately addressing a general case that provides the foundation needed to understand and critically compare different methods recommended for the apparent power resolution. This chapter emphasizes the need for the separation of the fundamental (50/60 Hz) active and reactive powers from the remaining apparent power components; this being an important IEEE Std. 1459–2010 contribution. Significant considerations are given to harmonics generation and injection mechanism, and how this phenomenon affects the power flow and the power definitions. It is shown how a certain amount of the fundamental power supplied to a nonlinear load is converted by nonlinear loads in a set of higher frequency components (harmonics) which, in turn, are injected into the power system. Flow charts that describe the flow path of different instantaneous power components are explained and used to demonstrate that the flow of instantaneous powers replicate the flow of Poynting vector components, thus providing the essential background for a correct decision when one needs to sort and define the apparent power components.

Chapter 4 is meant to provide the foundation needed for understanding approaches to power definitions, features, and limitations. The content is focused on apparent power resolutions as advocated by the researchers who directly, or indirectly, influenced the outcome of the two standards. They are:

C. I. Budeanu (1927)

S. Fryze (1932)

F. Buchholz (1950)

M. Depenbrock (1960)

L. S. Czarnecki (1984)

The author (1995)

The different methods are evaluated and compared by means of numerical examples. The transition to three-phase systems starts in Chapter 5. The scope is limited to sinusoidal balanced and unbalanced conditions. The key concept of Buchholz-Goodhue effective apparent power, the pivotal quantity of IEEE Std. 1459–2010, is introduced. Plenty of space is dedicated to the presentation of Fryze-Buchholz-Depenbrock (the FBD) method, which is the backbone of DIN 40110–2. The issue of power factor is discussed in detail. Ample explanations help prove the vector apparent power–the most popular apparent power definition–as being deficient. The mechanism of negative- and zero-sequence power generation is explained. A set of numerical examples allow the comparison of the definitions promoted by the DIN and the IEEE standard. It is concluded that for practical conditions, encountered in actual power systems, the two standards, while having different outcomes as concerns the zero-sequence voltage, yield very close results.

The sixth chapter deals with the most general case, the three-phase system with nonsinusoidal and unbalanced conditions. The literature abounds with studies of such conditions, promoting definitions that complete or contradict each other. This chapter presents only the most popular power definitions, i.e., those in use by instrumentation manufacturers in response to the request of electric utilities, or recommended by the two major standards. A central issue is debated: the IEEE Std. 1459–2010 considers that, ideally, a well compensated three-phase load must operate with positive-sequence currents and zero reactive power (unless the load is used as a power conditioner). DIN 40110–2 is different: it requires for unity power factor currents with waveforms that are replicas of the line–to an artificial neutral point voltage, (the actual neutral line is treated as the fourth phase). This means that traces of zero- and negative-sequence currents remain after the compensation. It is also explained that if all the loads supplied by a substation are compensated using either one of the two methods described in the two standards, the final results will be identical: perfect, pristine, positive-sequence sinusoidal currents with the respective symmetrical sinusoidal voltages.

The seventh chapter presents a newer, nonactive power, the randomness power. In situations when the monitored load is randomly time varying and power measurements are taken over a relatively large total observation time, the observation time is divided into small subintervals. The measurement taken for each subinterval is characterized by its own set of active and nonactive powers, stored for every subinterval. It is proved that the equivalent values for active and nonactive powers, measured over the total observation time, are the mean values of the active powers measured for each subinterval. This feature, however, does not apply to the measurements of apparent power, and it is necessary to include a randomness power in the resolution of apparent power, even when we deal with purely resistive loads. This new quantity gains in significance when the monitored loads are arc furnaces, welders, elevators, or any type of aleatory loads.

The last chapter includes eight appendices. The presented material is meant to help clarify heavier mathematical aspects, learn more about the Poynting vector applications, and find useful information that reveals the beauty of electromagnetic fields theory and the usefulness of Poynting vector in visualizing the energy flow. The reader interested in more rigorous mathematical demonstrations finds application of Lagrange multipliers to the computation of maximum active power. Another appendix deals with the computation of the allocated power loss to a monitored load connected in a network with a multitude of loads. Such information is crucial for the apparent power definition. The final appendix lists the readings of varmeters in the presence of distorted voltages and current waveforms.

I want to express my gratitude to Mrs. Catherine Emmerton, who coached my first steps in the complex world of LaTeX. Special appreciation goes to Prof. David Cyganski and Mr. Robert Brown for taking time from their busy schedule and bailing me out every time when "my PC was getting in trouble." Their friendship and camaraderie helped me to overcome some difficult moments during the preparation of the manuscript. It is my pleasant duty to acknowledge the diligent work of Dr. Grazia Todeschini, who was the first person to read the completed manuscript, flag typos and technical errors, and suggest improvements. A special thank you is extended to the team at Wiley including Simone Taylor, Nicky Skinner, Laura Bell and Clarissa Lim for their support in publishing this book. I would also like to thank Jane Utting for copyediting the book and also to Shalini Sharma at Aptara for typesetting the book. I am also indebted to many of my students who studied this material with me and inspired me by asking tough questions. Finally, my deepest appreciation goes to my beloved WPI, and to my colleagues who create and maintain an environment conducive to creativity and true camaraderie.

It is my sincere hope that this book will inspire and motivate the engineers and scientists that will design and build the new generation of smart meters to conform with the recommendations of the IEEE Std. 1459–2010 or the DIN 40110–2, and to continue the quest for more correct and practical apparent power definitions, symbolic mathematical expressions that will be embraced by all the electrical engineers living and working on all four corners of our Earth.

Alexander Eigeles Emanuel
Southborough, Massachusetts
January 2010

1

Electric Energy Flow: Physical Mechanisms

Through valuation only is there value; and without valuation the nut of existence would be hollow.
Hear it, ye creating ones!

—Fr. Nietzsche, *Thus Spake Zarathustra*

There are two schools of thought that help students visualize the flow of electric energy from source to load and grasp the basic relations among voltage, current, power, and energy. The first, and seemingly the simplest, explanation relies on the flow of electric charges represented in Fig. 1.1. We imagine a cylindrical conductor with a cross-sectional area A and length ℓ, containing uniformly distributed charged particles that carry a total electric charge q. The volume charge density is

$$\rho_v = \frac{q}{A\ell} \quad (\text{C/m}^3) \tag{1.1}$$

When a voltage v is applied between the ends of the conductive cylinder, a uniform electric field

$$E = \frac{v}{\ell} \quad (\text{V/m}) \tag{1.2}$$

is created within the conductor. The vector of this field is oriented parallel with the conductor. The interaction between the charged particles and the field E is causing their motion along the conductor. The force developed on the charged particles found within a thin slice of thickness dx, that holds the charge $dq = \rho_v A dx$ is $dF = E dq$. The total force applied on the entire charge held by the cylinder is

$$F = qE = A\ell\rho_v E = A\rho_v v \quad (\text{N}) \tag{1.3}$$

Once this system reaches steady-state the voltage source will pump continuously a constant flow of charge in a closed loop. One may picture this flow as the effect of a mechanical pressure

---

Power Definitions and the Physical Mechanism of Power Flow   Alexander Eigeles Emanuel
© 2010 John Wiley & Sons, Ltd

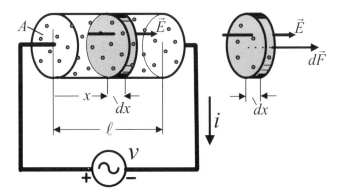

Figure 1.1   Flow of uniformly distributed charges in a homogeneous conductor.

$F/A = \ell\rho_v E = \rho_v v$. This model leads us straight to the notion of work or energy. To slide the total charge q a distance dx, consequent to the application of the force F, it is tantamount with doing the work

$$dw = Fdx = A\rho_v v dx \quad \text{(J)} \tag{1.4}$$

It may be assumed that the charged particles move with an average drift velocity $u = dx/dt$, proportional to the magnitude of the electric field, thus

$$u = KE \quad \text{(m/s)} \tag{1.5}$$

where the constant K is known as the mobility of the particles, (m^2/Vs). The elementary work dw is proportional to the drift velocity u. This fact becomes evident when (1.4) is written in the form

$$dw = A\rho_v v \frac{dx}{dt} dt = A\rho_v vu dt \tag{1.6}$$

The drift velocity $u = dx/dt$ is also hidden in the electric current expression

$$i = \frac{dq}{dt} = \frac{\rho_v A dx}{dt} = \rho_v Au \quad \text{(A)} \tag{1.7}$$

Substitution of (1.7) in (1.6) gives

$$dw = vi dt \tag{1.8}$$

During a time interval $t = t_2 - t_1$ the voltage source will generate the total energy

$$w = \int_{t_1}^{t_2} vi dt \quad \text{(J or Ws)} \tag{1.9}$$

The rate of flow of the electric energy at a particular time is the electric power

$$p = \frac{dw}{dt} = vi \quad (\text{W}) \tag{1.10}$$

From (1.7) and (1.5) we also obtain a simple deduction of Ohm's law. The current

$$i = \rho_v AKE = \frac{\rho_v K A}{\ell} E\ell = \frac{v}{R} \tag{1.11}$$

where

$$R = \frac{\ell}{\kappa A} \tag{1.12}$$

is the resistance of the conductor of length ℓ and cross-sectional area A, and $\kappa = \rho_v K$ is the specific conductivity of the observed conductive medium, $(\Omega m)^{-1}$.

Finally, equations (1.10) and (1.11) lead to the well known expressions of electric power

$$p = Ri^2 = \frac{v^2}{R} = vi \tag{1.13}$$

The above explanation of power and energy flow appears in some introductory textbooks of physics and is favored by electrical engineers that deal with low frequency equipment. A major drawback of this rudimentary model becomes apparent when we try to explain situations where the energy is stored in, or transferred through, dielectrics immersed in alternating electromagnetic fields, Fig. 1.2.

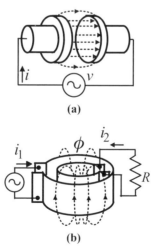

(a)

(b)

Figure 1.2 Examples where the Energy is Transferred via Dielectric Material: (a) Capacitor. (b) Magnetic Coupling.

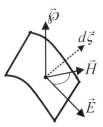

Figure 1.3 Poynting vector.

Engineers dealing with antennae, microwaves, and other high frequency applications long ago embraced a more advanced explanation based on a model loyal to the laws of storage, transmission, and dissipation of energy in any medium. This theory relies on the representation of the rate of flow of the energy density at any point in space by means of Poynting vector [1]

$$\vec{\wp} = \vec{E} \times \vec{H} \quad (\text{W/m}^2) \tag{1.14}$$

where $\vec{E}$ and $\vec{H}$ are the electric and magnetic field vectors at the considered point.

One will note that the unit W/m^2=Ws/m^{2}s discloses the fact that the Poynting vector quantifies the flow of electromagnetic energy per unit area per unit time, entering or exiting a defined point located on a virtual or actual surface, Fig. 1.3. Consequently the instantaneous power, entering or exiting a given surface ς, can be computed from the flux of the Poynting vector over the surface ς:

$$p = \int_{\varsigma} \vec{\wp} \cdot d\vec{\varsigma} \tag{1.15}$$

The Poynting vector approach was evaluated and improved by preeminent physicists like R. Feynman [2] and it is promoted in the best electromagnetic fields texts [3,4,5], being used as a most effective tool in the analysis and computations of eddy–currents, electromechanical torques, and electromagnetic energy radiation [6,7,8,9,10].

A few basic examples will help to reinforce the usefulness of this mathematical tool. First we consider a very long cylindrical conductor with radius a. We focus on a segment with length ℓ, Fig. 1.4. The voltage drop across the segment ℓ is v. We will use cylindrical coordinates $(x, \ r, \ \theta)$ with the versors $\vec{1}_x, \ \vec{1}_r, \ \vec{1}_\theta$. At any chosen point $(x, \ a, \ \theta)$, on the conductor's surface $2\pi a\ell$, Fig. 1.4a, the electric and magnetic field vectors are as follows:

$$\vec{E} = \frac{v}{\ell}\vec{1}_x \tag{1.16}$$

and

$$\vec{H} = \frac{i}{2\pi a}\vec{1}_\theta \tag{1.17}$$

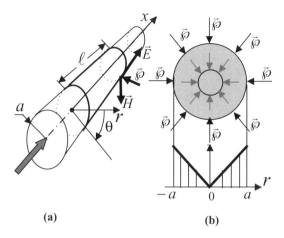

Figure 1.4 Poynting vector at the surface of a cylindrical conductor: (a) Three-dimensional sketch. (b) Cross-section.

Substitution of (1.16) and (1.17) in (1.14) gives the Poynting vector at the conductor's surface

$$\vec{\wp} = \vec{E} \times \vec{H} = -\frac{vi}{2\pi a \ell}\vec{1}_r \qquad (1.18)$$

This vector is perpendicular to the conductor's surface and is oriented toward conductor's center, Fig. 1.4b. If one reverses the direction of the current, both $\vec{E}$ and $\vec{H}$ will reverse direction, but $\vec{\wp}$ will keep its orientation unchanged, pointing toward the center of the conductor and carrying energy into the conductor. The power entering the conductor through the external surface $2\pi a \ell$ is

$$p = \int_{\varsigma} \vec{\wp} \cdot d\vec{\varsigma} = \wp 2\pi a \ell = \frac{vi}{2\pi a \ell} 2\pi a \ell = vi = Ri^2 \qquad (1.19)$$

This result sustains the concept that the conductor's Joule loss is provided by the energy impinged by the electromagnetic wave that enters the conductor's surface. Inside the conductor at any point on a cylindrical surface of radius $r \leq a$, the electric field vector (1.16) remains unchanged. The magnetic field vector, however, has an expression different than (1.17). Assuming uniform current density (a correct assumption only when the skin effect is negligible), the cylinder of radius $r \leq a$ encloses the current $(r/a)^2 i$, hence the magnetic field is

$$\vec{H} = \frac{\pi r^2}{\pi a^2}\frac{i}{2\pi r}\vec{1}_\theta = \frac{ri}{2\pi a^2}\vec{1}_\theta \qquad (1.20)$$

and in this case the Poynting vector inside the conductor

$$\vec{\wp} = \vec{E} \times \vec{H} = -\frac{rvi}{2\pi a^2 \ell}\vec{1}_r ; \quad r \leq a \qquad (1.21)$$

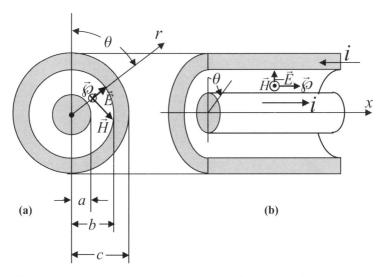

Figure 1.5   Coaxial cable: (a) Radial cross-section. (b) Axial cross-section.

remains radial and oriented toward the center. As the vector $\vec{\wp}$ nears the center, its magnitude is linearly decreasing, the density of the electromagnetic power is gradually diminished, and at $r = 0$ we find $\wp = 0$. The power impinged by the electromagnetic wave entering through the surface $2\pi r\ell$, of the internal cylinder of radius r, carries the power

$$p_r = 2\pi r \ell \wp = \left(\frac{r}{a}\right)^2 vi$$

that covers exactly the Joule loss within the volume $\pi r^2 \ell$.

In the next example we consider a coaxial cable with the radii a, b and c, Fig. 1.5. The electric and magnetic fields within the dielectric are

$$\vec{E} = \frac{v}{\ln(b/a)} \frac{1}{r} \vec{1}_r \qquad (1.22)$$

and

$$\vec{H} = \frac{i}{2\pi r} \vec{1}_\theta \qquad (1.23)$$

yielding the Poynting vector

$$\vec{\wp} = \vec{E} \times \vec{H} = \frac{1}{2\pi \ln(b/a)} \frac{vi}{r^2} \vec{1}_x \qquad (1.24)$$

parallel with the coaxial cable and oriented toward the load.

The electromagnetic power carried on the "wings" of the vector $\vec{\wp}$, through the cross-section $\pi(b^2 - a^2)$, is

$$p = \int_{r=a}^{b} \vec{\wp} \cdot \vec{1}_x \, 2\pi r \, dr = \frac{2\pi vi}{2\pi \ln(b/a)} \int_{r=a}^{b} \frac{dr}{r} = vi \qquad (1.25)$$

The last result emphasizes the fact that the flow of electric energy toward the load takes place within the dielectric that surrounds the transmission line conductors. One may figure the conductors as a wave-guide for the electromagnetic wave. Equation (1.24) shows that the density of the energy increases as one nears the conductors. In the vicinity of a superconductor $\vec{\wp}$ is perfectly parallel to the conductor; however, in the vicinity of a lossy conductor the Poynting vector streamlines bend slightly toward the conductor due to a small component perpendicular to the conductor surface. This transversal component of $\vec{\wp}$ transfers to the conductors the power that sustains the Joule and eddy-current losses dissipated in the conductor.

On the surface of magnetic cores the Poynting vector supplies hysteresis and eddy-current losses as well as energy stored in, and returned from, the magnetic field. Since part of the magnetic fields are distributed within conductors and semiconductors, the energy stored in the magnetic field located in the conductors is also transported by the $\vec{\wp}$ component perpendicular to the conductors' surface. The same observation holds true for dielectrics; both the flow of energy that is converted in heat due to dielectric losses and the flow of oscillating energy tied to the energy stored in the electric field, can be visualized and quantified with the help of $\vec{\wp}$ and its components.

The fans of the theory that advocates the concept of energy flow strictly confined to the conductor material found support in the Slepian vector theory [6,7]. The vector $\vec{\zeta}$ is a modified form of the Poynting vector where, besides the easily understood longitudinal flow of energy, expressed by the vector

$$\frac{p}{A}\vec{1}_x = \frac{vi}{A}\vec{1}_x = v\vec{j} \quad (\text{W/m}^2)$$

also the flow of energy stored in, or delivered by, the electric and magnetic fields are included, leading to the following expression

$$\vec{\zeta} = v\vec{j} + v\frac{\partial \vec{D}}{\partial t} + \vec{H} \times \frac{\partial \vec{A}}{\partial t}$$

where
  $\vec{j} = i/A$ is current density vector (A/m^2),
  $\vec{D} = \epsilon \vec{E}$ is the electric flux density vector (C/m^2) and
  $\vec{A}$ is the magnetic vector potential (Vs/m).

The Poynting vector approach seems simpler and more pliable than the Slepian vector. The Poynting vector theory provides an excellent tool that helps to comprehend the details of the physical mechanism of electric energy flow, the separation of electric power in diverse active and nonactive components, and, most importantly, it helps to evaluate the merits of definitions used to quantify the quality of electric energy.

More information on Poynting vector is provided in Chapter 8, Appendices I, II, III and IV.

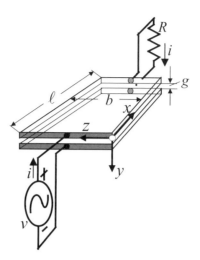

Figure 1.6   Rectangular and parallel superconductors connecting a voltage source with a load.

ADDITIONAL EXAMPLES

1. In Fig. 1.6 are sketched two parallel rectangular superconductors—with length ℓ and width b—separated by a small dielectric gap $g << b$. The voltage source v supplies to the load R with the current i. The electric field vector, assumed uniform (fringing effects are neglected), is

$$\vec{E} = \frac{v}{g}\vec{1}_y$$

and the magnetic field in the dielectric space $0 < x < \ell, 0 < y < g$ and $0 < z < b$ is

$$\vec{H} = \frac{i}{b}\vec{1}_z$$

The Poynting vector is now readily obtained,

$$\vec{\wp} = \vec{E} \times \vec{H} = \frac{vi}{bg}\vec{1}_x \;; \quad (\vec{1}_y \times \vec{1}_z = \vec{1}_x)$$

As expected we note that $\vec{\wp}$ is oriented toward the load and its flux through the cross-section area bg gives the power

$$p = \wp bg = vi$$

Since we assumed superconductors, the electric field within the conductors is nil and no Poynting vector component in the y-direction exists.

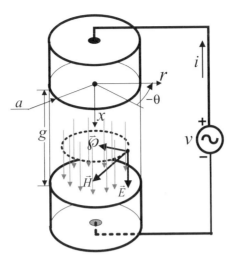

Figure 1.7   Capacitor with parallel circular electrodes.

2. A capacitor with two circular parallel electrodes of radius a spaced apart at the distance g, Fig. 1.7, will help us to understand how the Poynting vector is used to analyze systems where the energy is transferred via a material with dielectric permitivity ϵ.

The current flowing through the capacitance C is a displacement current

$$i = C\frac{dv}{dt}; \quad C = \epsilon\frac{\pi a^2}{g}$$

whose density vector is

$$\vec{j} = \frac{i}{\pi a^2}\vec{1}_x = \frac{\epsilon}{g}\frac{dv}{dt}\vec{1}_x \quad (A/m^2)$$

To find the magnetic field $\vec{H}$ we will apply Ampère's law to a circle of radius $r < a$, concentric with the electrodes. The Ampère's law in its integral form is

$$\oint_C \vec{H}\cdot d\vec{\ell} = \int_s \vec{j}\cdot d\vec{s}$$

meaning that the closed contour line integration of $\vec{H}$ along the perimeter of the circle r yields a value equal to the current flowing through the circle. Since this system is symmetrical the vector $\vec{H}$ is tangent to the circle r, its magnitude along the circle r is constant and its radial and axial components are nil. The Ampère's law can now be written as follows:

$$\int_{\theta=0}^{2\pi} Hr\,d\theta = \int_{r=0}^{r} j\,2\pi r\,dr$$

that gives

$$2\pi r H = j\pi r^2$$

and from here follows

$$\vec{H} = \frac{r}{2} j \, \vec{1}_\theta \quad \text{for } 0 \le r \le a$$

We recognize that the electric field vector is controlled by the voltage v and the gap g, hence

$$\vec{E} = \frac{v}{g} \, \vec{1}_x$$

and the Poynting vector is now easily obtained

$$\vec{\wp} = \vec{E} \times \vec{H} = \left(\frac{v}{g} \, \vec{1}_x \right) \times \left(\frac{r\epsilon}{2g} \frac{dv}{dt} \, \vec{1}_\theta \right) = \frac{-v r \epsilon}{2g^2} \frac{dv}{dt} \, \vec{1}_r$$

$$= \frac{-r\epsilon}{2g^2} \frac{1}{2} \frac{d}{dt} (v^2) \, \vec{1}_r \, ; \quad (\vec{1}_x \times \vec{1}_\theta = -\vec{1}_r)$$

We learn that the Poynting vector entering or leaving the dielectric has a radial distribution. As it penetrates the dielectric its intensity linearly decreases and at $r = 0$ $\vec{\wp} = 0$. The power that enters or exits the external surface of the dielectric, $2\pi a g$ is

$$P_{r=a} = \int_{x=0}^{g} \vec{\wp} \cdot (2\pi a \, dx) \, \vec{1}_r = -\frac{a\epsilon}{2g^2} \frac{1}{2} \frac{d(v^2)}{dt} 2\pi a g = -\frac{1}{2} \epsilon \frac{\pi a^2}{g} \frac{d(v^2)}{dt} = -\frac{1}{2} C \frac{d(v^2)}{dt}$$

We found that this power is just the rate of flow of the energy stored in the capacitor,

$$p = -\frac{d}{dt} \left(\frac{1}{2} C v^2 \right) = -\frac{dw_e}{dt} = -Cv \frac{dv}{dt}$$

If we assume a sinusoidal voltage

$$v = \sqrt{2} V \sin \omega t$$

we obtain

$$p = -Cv \frac{dv}{dt} = -C[\sqrt{2} V \sin(\omega t)][\sqrt{2} \omega V \cos(\omega t)]$$

$$= -C \omega V^2 \sin(2\omega t)$$

The power p varies sinusoidally in time with twice the angular frequency ω. The mean value of the power is nil. Energy enters and exits the dielectric. For a quarter of cycle, $T/4 = \pi/2\omega$, the energy exits the dielectric gradually returning to the source, and in the next quarter of cycle the energy flow reverses and the oscillations continue.

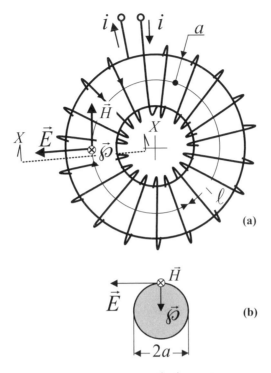

Figure 1.8　Toroidal inductor: (a) Core, winding, and $\vec{E}$, $\vec{H}$ and $\vec{\wp}$ vectors. (b) Cross-section XX and the $\vec{E}$, $\vec{H}$, and $\vec{\wp}$ vectors.

3. A toroidal inductor, Fig. 1.8a, has a ferromagnetic core with a mean path length ℓ, circular cross-sectional area with radius a and permeability μ (H/m) and carries a winding with N turns uniformly distributed along the core's surface. The winding resistance is very small when compared with the reactance and can be ignored. Through the winding flows a current $i = i(t)$ producing a magnetic field

$$H \approx Ni/\ell \quad \text{(A/m)}$$

The magnetic flux within the core is

$$\phi = \pi a^2 \mu Ni/\ell \quad \text{(Vs)}$$

Along the circular perimeter $2\pi a$, Fig. 1.8b, the induced voltage $d\phi/dt$ is sustained by the electric field

$$E = \frac{d\phi/dt}{2\pi a} = \frac{\mu Na}{2\ell}\frac{di}{dt} \quad \text{(V/m)} \tag{1.26}$$

This electric field is tangential to the core surface. Since the vectors $\vec{E}$ and $\vec{H}$ are perpendicular, the resulting Poynting vector is perpendicular to the core surface, Fig. 1.8b, and is uniformly distributed over the entire $2\pi a\ell$ surface. Its magnitude is

$$\wp = |\vec{E} \times \vec{H}| = \mu \frac{N^2 a}{2\ell^2} i \frac{di}{dt}$$

The total power impinged into the core by $\vec{\wp}$ is

$$p = 2\pi a\ell\,\wp = \mu \frac{\pi a^2 N^2}{\ell} i \frac{di}{dt} = Li \frac{di}{dt}$$

where

$$L = \mu \frac{\pi a^2 N^2}{\ell} \quad \text{(H)}$$

is the inductance.

Assuming the current $i = \sqrt{2}I \sin(\omega t)$ we find the voltage $v = Ldi/dt = \sqrt{2}\omega LI \cos(\omega t)$ and the resulting instantaneous power is

$$p = vi = 2\omega LI^2 \sin(2\omega t)$$

Returning to the Poynting vector expression for this case we find

$$\wp = \mu \frac{N^2 a}{2\ell^2} i \frac{di}{dt} = \mu \frac{N^2 a}{2\ell^2} \sqrt{2}I \sin(\omega t)\omega\sqrt{2}I \cos(\omega t) = \frac{\omega L}{\pi a\ell} I^2 \sin(2\omega t) = \frac{p}{2\pi a\ell}$$

thus

$$p = 2\pi a\ell\,\wp = 2\omega LI^2 \sin(2\omega t)$$

meaning that the power is transmitted from the winding to the magnetic core through the dielectric materials that insulates the conductor from the core.

All the above examples show that $\wp$ is an instantaneous power density, W/m^2, that has active and nonactive components. In the first example there is a unidirectional transport of energy from the source to the load and the load, a resistance, absorbs and dissipates active power. In the next two examples the loads did not convert the energy in heat and the energy is oscillating back and forth between the source and the loads. No net transfer of energy takes place and in these cases we deal with a nonactive type of power. In all three examples the Poynting vector $\wp$ helps to obtain the visual picture of the energy flow from the source to the load through the space surrounding the conductors, giving complete information about the power's time variation and space distribution [12].

4. We will look now into a case where both active and nonactive powers exist. We repeat the previous example with a toroidal inductor of identical geometry, Fig. 1.9a, but with the ferromagnetic core made of a material that has an idealized hysteresis loop, Fig.1.9b. The voltage impressed across this nonlinear inductor is adjusted to maintain the magnetic induction excursion from $-B_S$ to B_S, during the positive half-cycle, and back during the negative half-cycle, Fig. 1.9c. During the positive half-cycle the current increases from $i = 0$

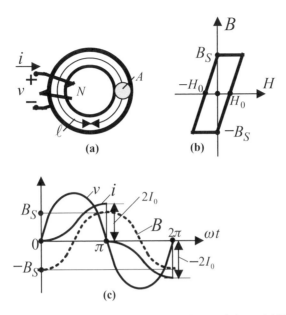

Figure 1.9 Toroidal nonlinear inductor: (a) Geometry. (b) Hysteresis loop. (c) Voltage, induction, and current waveforms.

to $i = 2I_0 = 2H_0\ell/N$, where H_0 is the coercive magnetic field. The time variation of the magnetic induction B is governed by Faraday's law, and assuming a sinusoidal input voltage we have

$$\sqrt{2}V \sin(\omega t) = NA\frac{dB}{dt} \tag{1.27}$$

where $A = \pi a^2$ is the cross-sectional area of the toroid. The integration of (1.27) gives

$$B = K - \frac{\sqrt{2}V}{NA\omega}\cos(\omega t)$$

at $\omega t = 0$ the core is in negative saturation, $B = -B_S$, therefore

$$-B_S = K - \frac{\sqrt{2}V}{NA\,\omega}$$

and the integration constant is

$$K = -B_S + \frac{\sqrt{2}V}{NA\,\omega}$$

hence the induction varies in time according to the expression

$$B = -B_S + \frac{\sqrt{2}V}{NA\omega}[1 - \cos(\omega t)] \tag{1.28}$$

At $\omega t = \pi$ the core reaches positive saturation, $B = B_S$, thus

$$B_S = -B_S + \frac{\sqrt{2}V}{NA\omega}[1 - \cos(\pi)]$$

and from here we find

$$B_S = \frac{\sqrt{2}V}{NA\omega}$$

This result enables us to rewrite (1.28) in a more convenient form

$$B = -B_S + B_S[1 - \cos(\omega t)] = -B_S\cos(\omega t) \quad \text{for } 0 < \omega t < \pi \tag{1.29}$$

The magnetic field H, during the positive half-cycle, Fig. 1.9b, has the equation

$$H = H_0 + \frac{H_0}{B_S}B \tag{1.30}$$

Substitution of (1.29) in (1.30) yields

$$H = H_0[1 - \cos(\omega t)] \tag{1.31}$$

Since $i = H\ell/N$, from (1.30) results

$$i = I_0[1 - \cos(\omega t)] \quad \text{for } 0 < \omega t < \pi \quad \text{with } I_0 = H_0\ell/N$$

and in the same way it is found

$$i = -I_0[1 + \cos(\omega t) - \pi] \quad \text{for } \pi < \omega t < 2\pi$$

This nonsinusoidal current, Fig. 1.9c, has the following Fourier series:

$$i = \frac{4I_0}{\pi}\sin(\omega t) - I_0\cos(\omega t) + \frac{4I_0}{\pi}\left[\frac{1}{3}\sin(3\omega t) + \frac{1}{5}\sin(5\omega t) + \cdots\right] \tag{1.32}$$

The active instantaneous power supplied to the load is given by the product vi_p, where

$$i_p = (4I_0/\pi)\sin(\omega t)$$

is the current component in-phase with the voltage. The active power is the product of the rms voltage with the rms active current, i.e.

$$P = \frac{4I_0 V}{\pi\sqrt{2}} = \frac{2\sqrt{2}}{\pi} V I_0 \quad (W)$$

Now since the winding resistance was ignored this active power P is due to hysteresis losses only. This affirmation is proved if we compute the hysteresis losses straightforward with the formula

$$P_{Hyst} = (HystLoopArea)(CoreVolume)f$$

where
  $HystLoopArea = 4B_S H_0$ (Ws/m^3)
  $CoreVolume = A\ell$ (m^3)
  $f = \omega/2\pi$ is the frequency, (1/s), thus

$$P_{Hyst} = (4B_S H_0)(A\ell)f = 4\frac{\sqrt{2}V}{AN\omega}\frac{N I_0}{\ell}\frac{A\ell\omega}{2\pi} = \frac{2\sqrt{2}}{\pi}V I_0 = P$$

At this point we shall bring in the picture the Poynting vector $\vec{\wp}$. At the surface of the toroidal core the electric field vector $\vec{E}$ is tangential to the surface, flowing along circles of radius a (tangential to the perimeter of the core cross-section, Fig.1.8b). The field E is given by (1.26)

$$E = \frac{d\phi/dt}{2\pi a} = \frac{1}{2\pi a}\frac{d}{dt}\left(\frac{-\sqrt{2}V}{\omega N}\cos(\omega t)\right) = \frac{\sqrt{2}V}{2\pi a N}\sin(\omega t)$$

The magnetic field vector $\vec{H}$ is also tangential to the toroidal core surface, but perpendicular to $\vec{E}$. The vector $\vec{H}$ flows along circles with the centers located on the toroid's axis.
From (1.32) we find the following equation for the magnetic field

$$H = \frac{Ni}{\ell} = \frac{4I_0 N}{\pi\ell}\left[\sin(\omega t) - \frac{\pi}{4}\cos(\omega t) + \sum_{h=3,5,7,\cdots}\frac{1}{h}\sin(h\omega t)\right] \qquad (1.33)$$

The Poynting vector is perpendicular to the toroid's surface and has three distinctive terms derived from (1.33)

$$\wp = \wp_p + \wp_q + \wp_H$$

where

$$\wp_p = \left[\frac{\sqrt{2}V}{2\pi a N}\sin(\omega t)\right]\left[\frac{4N I_0}{\pi\ell}\sin(\omega t)\right] = \frac{2\sqrt{2}V I_0}{\pi^2 a\ell}\sin^2(\omega t)$$

is the active component of $\vec{\wp}$. The instantaneous power impinged in the core is

$$p_p = 2\pi a\ell \frac{2\sqrt{2}V I_0}{\pi^2 a\ell} \sin^2(\omega t)$$

and its mean value, the active power is

$$P = \frac{2\sqrt{2}V I_0}{\pi} = P_{Hyst}$$

The second term

$$\wp_q = \left[\frac{\sqrt{2}V}{2\pi a N} \sin(\omega t)\right]\left[\frac{N I_0}{\ell} \cos(\omega t)\right] = \frac{\sqrt{2}V I_0}{4\pi a\ell} \sin(2\omega t)$$

is the reactive component; it oscillates in and out of the core without causing power loss in the core. It is due to the energy accumulated in the magnetic field and returned back to the source.

The third component

$$\wp_H = \left[\frac{\sqrt{2}V}{2\pi a N} \sin(\omega t)\right]\frac{4 N I_0}{\pi\ell} \sum_{h=3,5,7,\cdots} \frac{1}{h}\sin(h\omega t)$$

is due to the current harmonics. It is a nonactive component that contains many subcomponents that also oscillate to and fro, from source to the inductor and back, in and out of the core with the frequencies $(h\pm 1)\omega$, $h = 3, 5, \cdots$. Both $\wp_q$ and $\wp_H$ have similar electromagnetic "signatures" that result from the $\vec{E}\times\vec{H}$ interaction.

1.1   Problems

1.1 Show that the Poynting vector at the surface $2\pi b\ell$, Fig. 1.5, transports the energy dissipated in the volume of the external conductor of radii b and c.

1.2 A superconductor with a length ℓ and rectangular cross-section $g \times b$, slides frictionless in an air-gap g where a uniform magnetic field H_o was established, Fig. 1.10. A current i is injected through the conductor. A Lorenz force

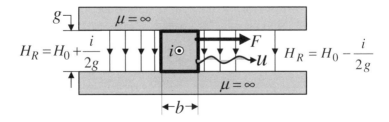

Figure 1.10   Rectangular conductor moving in a perpendicular magnetic field.

$$F = \mu_0 H_0 \ell i$$

is developed and the conductor moves toward the right with a velocity u. A motional voltage, or back EMF, is induced in the moving bar. Since we assume a superconductor, this back EMF equals the voltage v impressed across the conductor, i.e.

$$v = \mu_0 H_0 \ell u$$

When this electromechanical system reaches steady-state the electric field at the surface of the conductor is

$$E = \frac{v}{\ell} = \mu_0 H_0 u$$

On the lateral faces of the conductor the magnetic field vector has two vertical components; first is H_0 and the second $\pm i/2g$ is the magnetic field produced by the current i.

Using the Poynting vector, compute the electromechanical power supplied to this superconductor and prove that

$$p = \mu_0 H_0 u \ell i = vi = Fu$$

Repeat the same analysis for generator operation. Assume that the motion is maintained toward the right, but the flow of the current and the direction of $\vec{E}$ are reversed on the account of the motional voltage.

1.3 A cylindrical solenoid, radius a, length g and N turns, is clamped in a concentrical ferromagnetic frame, Fig. 1.11, with $\mu \to \infty$. Assuming $g \ll a$, show by means of Poynting vector that the power entering the envelope $2\pi a g$ is

$$p = \frac{d}{dt}\left(\frac{Li^2}{2}\right); \quad L = \frac{\mu_0 N^2 \pi a^2}{g}$$

1.4 Given a superconductive coaxial cable with radii $a < b$ and length ℓ that carries a current i, show by means of Poynting vector that the flux of the powers entering the surfaces $2\pi a \ell$ and $2\pi b \ell$ carries a total power

$$p = \frac{d}{dt}\frac{1}{2}Li^2; \quad L = \frac{\mu_0 \ell}{2\pi}\ln\frac{b}{a}$$

1.5 A superconductive coaxial cable (length ℓ radii a and b, see Fig. 1.5) has insulation made of a dielectric with the permittivity ϵ. One end is supplied with the voltage $v = v(t)$, the other end of the cable is left open (disconnected).

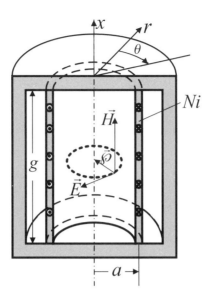

Figure 1.11   Cylindrical solenoid mounted in a ferromagnetic frame: Cross-sectional view.

Prove that the Poynting vector entering the area $\pi(b^2 - a^2)$ at the supplying end carries the power

$$p = \frac{d}{dt}\left(\frac{1}{2}Cv^2\right) \quad \text{where } C = \frac{2\pi\epsilon\ell}{\ln\left(\frac{b}{a}\right)}$$

Also show that as the Poynting vector enters the dielectric, it loses intensity and at the opposed end, where $x = \ell$, $P = 0$.

1.2   References

[1] Poynting J. H.: "On the Transfer of Energy in the Electromagnetic Field," Philosophical Transactions of the Royal Society, 1884, p. 175.

[2] Feynman R. P., Leighton R. B., Sand M. L.: *"The Feynman Lectures on Physics,"* Addison-Wesley, 1964, Vol. II, Chap. 27, pp. 1–11.

[3] Demarest K. R.: *"Engineering Electromagnetics,"* Prentice Hall, 1988, p. 462.

[4] Sadiku M. N. O.: *"Elements of Electromagnetism,"* Sounders College Publishing, 1989, p. 457.

[5] Haus H. A. and Melcher J. R.: *"Electromagnetic Fields and Energy,"* Prentice Hall, 1988, p. 462.

[6] Slepian J.: "The Flow of Power in Electrical Machines," Electrical Journal, Vol. 16, 1919, pp. 303–11.

[7] Slepian J.: "Energy and Energy Flow in the Electromagnetic Field," Journal of Applied Physics, Vol. 13, Aug. 1942, pp. 512–18.

[8] Ferreira J. A.: "Application of the Poynting Vector for Power Conditioning and Conversion," IEEE Transactions on Education, Vol. 31, No. 4, Nov. 1988, pp. 257–64.

[9] Hawthorne E. I.: "Flow of Energy in Synchronous Machines," AIEE Transactions, Vol. 73, March 1954, pp. 1–9.

[10] Ferreira J. A., Van Wyk J. D.: "Electromagnetic Energy Propagation in Power Electronic Converters: Toward Future Electromagnetic Integration," Proc. of IEEE, Vol. 89, No. 6, June 2001, pp. 876–89.
[11] Emanuel A. E.: "Poynting Vector and the Physical Meaning of Nonactive Powers," IEEE Transactions on Instrumentation and Measurement, Vol. 54, No. 4, August 2005, pp. 1457–62.
[12] Emanuel A. E.: "About the Rejection of Poynting Vector in Power Systems Analysis," Electrical Power Quality and Utilization Journal, Vol. XIII, No. 1, 2007, pp. 41–47.

2

Single-Phase Systems With Sinusoidal Waveforms

> Those who can not remember the past are condemned to repeat it.
>
> —George Santayana

This chapter examines the steady-state flow of electric energy into two terminal loads supplied with sinusoidal voltage. It starts with the basic components, the linear R, L and C. Their study provides the bricks and mortar needed to build the steps that lead to the understanding of the controversial polyphase nonsinusoidal systems. The scope of this chapter is to introduce modeling concepts and symbols used throughout this book. Some of the material is well covered in many textbooks, however, there are sections that provide new insights and help to explain the concept of active and nonactive power based on the electromagnetic field theory and to detail the space and time distribution of the power flow.

2.1 The Resistance

We start with the investigation of the circuit shown in Fig. 2.1a. A sinusoidal voltage source

$$v = \widehat{V} \sin(\omega t) \tag{2.1}$$

is impressed across the resistance R causing the flow of a sinusoidal current in-phase with the voltage,

$$i = \frac{v}{R} = \widehat{I} \sin(\omega t) \tag{2.2}$$

where
$\widehat{V} = \sqrt{2}V$ is the amplitude and V the rms value of the voltage v,
$\omega = 2\pi f$ is the angular frequency (rad/s), f is the frequency (Hz) and $T = 1/f$ is the period (s).

Power Definitions and the Physical Mechanism of Power Flow   Alexander Eigeles Emanuel
© 2010 John Wiley & Sons, Ltd

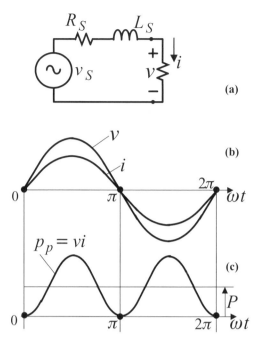

Figure 2.1 Resistance supplied by a sinusoidal voltage: (a) Circuit. (b) Load voltage and current waveform. (c) Instantaneous power waveform.

$\widehat{I} = \widehat{V}/R; \widehat{I} = \sqrt{2}I$ is the amplitude and $I = V/R$ is the rms value of the current i.

The instantaneous power p_p supplied to the load R is obtained by multiplying the instantaneous voltage v and the current i,

$$p_p = vi = \widehat{V}\widehat{I}\sin^2(\omega t) = \frac{1 - \cos(2\omega t)}{2} \, \widehat{V}\widehat{I}$$

$$= \frac{\widehat{V}\widehat{I}}{2} - \frac{\widehat{V}\widehat{I}}{2}\cos(2\omega t) = VI - VI\cos(2\omega t) \qquad (2.3)$$

This power is not constant in time, it is a cosinusoidal oscillation with an amplitude VI and a frequency twice the voltage source frequency, Fig. 2.1c. It oscillates between 0 and $2VI$. The mean value VI of the instantaneous power over one cycle $T = 2\pi/\omega$ can be easily found by inspecting (2.3), nevertheless we will refer to the basic definition

$$P = \frac{1}{T}\int_0^T p_p \, dt = \frac{1}{T}\int_0^T vi \, dt \qquad (2.4)$$

Substitution of (2.1) and (2.2) in (2.4) gives

$$P = \frac{1}{T} \int_0^T \widehat{V}\widehat{I}\sin^2(\omega t)\, dt = \frac{1}{2\pi} \int_0^{2\pi} \widehat{V}\widehat{I}\sin^2(\omega t)\, d(\omega t) = \frac{\widehat{V}\widehat{I}}{2} = VI \qquad (2.5)$$

Equation (2.3) can be also written in the form

$$p_p = P - p_i = P - P\cos(2\omega t) \qquad (2.6)$$

that reveals at a glance the structure of the rate of electric energy flow to a resistance supplied with sinusoidal voltage. The first term is the average power

$$P = VI = RI^2 = \frac{V^2}{R}$$

known as the *active power* or *real power* (sometimes also called *effective power*) and it is measured in W (Watt). The second term, $p_i = -P\cos(\omega t)$ is the oscillation with amplitude P, and it is always present when a sinusoidal voltage supplies a resistance. This term is the *instantaneous intrinsic power*. Its physical significance becomes clearer if we focus on the flow of energy into the load R. During a time interval from 0 to t the amount of energy supplied to this load is

$$w_p = \int_0^t p_p\, dt = \int_0^t [P - P\cos(2\omega t)]\, dt = Pt - \frac{P}{2\omega}\sin(2\omega t) \qquad (2.7)$$

The first term, Pt, that steadily ramps up in time, Fig. 2.2, quantifies the net flow of energy delivered to R over an integer number of half-cycles. For example, the energy converted in heat during N half-cycles is

$$W = P(NT/2)$$

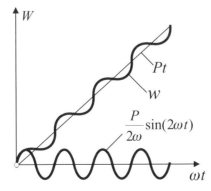

Figure 2.2 Time-variation of the energy and its components.

The second term, $[P \sin(2\omega t)]/2\omega$, makes no contribution to the net transfer of energy over the time $NT/2$; however, this term is always present when active power is delivered in ac systems. In a dc circuit, under steady-state conditions, the instantaneous power supplied to a resistor $p_p = RV^2 = V^2/R$ is perfectly constant, hence as long as R and V are constant no intrinsic power is present.

Note: One may find in the engineering literature [1] attempts to attach more significance to the oscillation $P \cos(2\omega t)$ and to single it as a consequential power component. Such a step is not warranted. The reason for opposing such a separation stems from the physical meaning of the rms value.

Let us assume that a resistance R converts in heat, during a limited time τ, the energy W. This means that the average power is

$$P = \frac{W}{\tau} \quad \text{with} \quad W = \int_0^{\tau} p \, dt$$

For the observation time τ, the rms current will be governed by expression

$$I = \sqrt{\frac{P}{R}} = \sqrt{\frac{W}{\tau R}}$$

From here results that when a given amount of energy W is dissipated by a resistance R over a time τ, there is only one value of rms current that fulfills this condition for the entire duration τ and this value, I, is independent of the instantaneous power fluctuations. If the resistance R is supplied by a sinusoidal voltage or by a perfect direct voltage—in which case there is not intrinsic instantaneous power—the same amount of energy will be delivered during the time τ as long as the rms currents are the same for both situations. For example, if we assume that the instantaneous power dissipated by a resistance has the expression

$$p_p = P + F_p(t)$$

where $F_p(t)$ is the fluctuating component with a nil average value, i.e.

$$\langle F_p(t) \rangle = \frac{1}{\tau} \int_0^{\tau} F_p(t) \, dt = 0$$

then the rms current $I = \sqrt{P/R}$ is independent of $F_p(t)$ and so is W. This observation is depicted in Fig. 2.3 where four different graphs representing time-variations $w = w(t)$ are shown. All four trajectories designate the delivery of the same amount of energy W over the time $\tau = NT/2$. The straight line A corresponds to a perfect dc condition that is considered ideal for energy transfer. The undulating graph B is for the sinusoidal case, and the graphs C and D are for two different hypothetical transient conditions. All four trajectories yield the same rms current.

In the process of electric energy transfer to a load the amount of energy lost in the transmission line is of great consequence. The economist wants to know the cost of the lost energy and the design engineer wants to determine the temperature of the conductors, evaluate the aging

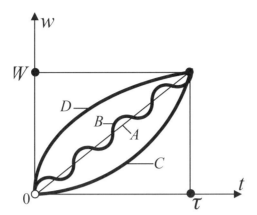

Figure 2.3 The same amount of energy W is delivered during the time τ. All four trajectories yield the same average power $P = W/\tau$.

of the cables, compute mechanical stresses and the eventual conductor sag. We find that the average power loss in the line resistance R_s, Fig. 2.1a, is

$$\Delta P_p = R_s I^2 = R_s \left(\frac{P}{V}\right)^2 \tag{2.8}$$

and in this particular case depends solely on the equivalent line resistance R_s, the active power P and the load rms voltage V.

2.2 The Inductance

Now the resistance R, of Fig. 2.1a, is replaced by an inductance L, Fig. 2.4a.

The cosinusoidal current

$$i = -\widehat{I}\cos(\omega t) \tag{2.9}$$

causes a voltage drop

$$v = L\frac{di}{dt} = \omega L \widehat{I}\sin(\omega t) = \widehat{V}\sin(\omega t); \quad \widehat{V} = \omega L\widehat{I}; \quad V = \omega L I \tag{2.10}$$

and the instantaneous power

$$p_{qL} = vi = -\widehat{V}\widehat{I}\sin(\omega t)\cos(\omega t) = -\frac{\widehat{V}\widehat{I}}{2}\sin(2\omega t) = -VI\sin(2\omega t) = -Q_L\sin(2\omega t) \tag{2.11}$$

where the amplitude of the instantaneous power oscillation is

$$Q_L = VI = \omega L I^2 = \frac{V^2}{\omega L}$$

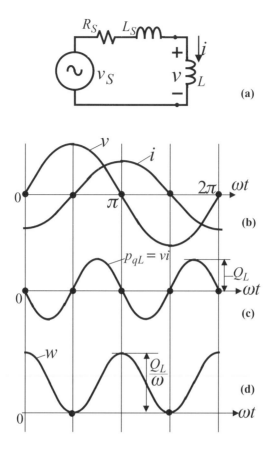

Figure 2.4 Inductance supplied by a sinusoidal voltage: (a) Circuit. (b) Inductance voltage and current waveforms. (c) Instantaneous power waveform. (d) Energy.

Since no electric energy is converted in heat or mechanical energy, there is no active power

$$P = \frac{1}{T} \int_0^T p_{qL} \, dt = \frac{1}{T} \int_0^T -Q \sin(2\omega t) \, dt = 0 \tag{2.12}$$

The energy characterized by the rate of flow p_{qL} has the following expression:

$$w_{qL} = \int p_{qL} \, dt = \int -Q_L \sin(2\omega t) \, dt = K + \frac{Q_L}{2\omega} \cos(2\omega t) \tag{2.13}$$

where K is an integration constant. In steady-state at $t = 0$ the current $i = -\widehat{I}$, Fig. 2.4b, and the energy stored in L is

$$w_{qL}|_{t=0} = \frac{L\widehat{I}^2}{2} = \frac{\omega L \widehat{I}^2}{2\omega} = \frac{\widehat{V}\widehat{I}}{2\omega} = \frac{Q_L}{\omega}$$

Thus

$$K = \left[w_{qL}|_{t=0} - \frac{Q_L}{2\omega} \cos(0) \right] = \frac{Q_L}{\omega} - \frac{Q_L}{2\omega} = \frac{Q_L}{2\omega}$$

and

$$w_{qL} = \frac{Q_L}{2\omega}[1 + \cos(2\omega t)] \qquad (2.14)$$

Typical oscillograms are presented in Figs. 2.4b, c and d. Under steady-state conditions the energy is continuously oscillating between L and the voltage source v_s. At the beginning of each voltage half-cycle, when the current is reaching peak value, the stored energy reaches the maximum value, Q_L/ω. During the next quarter-cycle the energy is gradually returned to the source v_s till the inductance is totally depleted of energy. In the following quarter-cycle the inductance is again charged to its fullest. This fluctuation of energy between L and v_s is sustained by the current i (with the rms value $I = V/\omega L = Q_L/V$.) This current is causing power loss in the line resistance R_s

$$\Delta P_Q = R_s I^2 = R_s \left(\frac{Q_L}{V} \right)^2 \qquad (2.15)$$

The amplitude Q_L of the instantaneous power oscillation is called *reactive power* and it is measured in var (volt-ampere-reactive). This reactive power Q_L belongs to the category of powers defined as *nonactive powers*.

These powers quantify the rate of flow of the energy exchanged among loads and sources, or even among different loads. Such exchanges of energy do not amount to a net transfer of energy between the source and the load, but are the sources of additional power loss. It is important to observe that both the intrinsic instantaneous power $p_i = -P \cos(2\omega t)$, (see (2.6)), and the reactive instantaneous power $p_{qL} = -Q_L \sin(2\omega t)$, are similar power oscillations with identical frequencies having no average value; however, p_i is always bound up to an active power and does not cause power loss, while p_{qL} does cause power loss in the conductors that supply the load.

2.3 The Capacitance

This case is similar to the previous one. The inductance L in Fig. 2.4 is replaced by the capacitance C. The cosinusoidal current

$$i = \widehat{I} \cos(\omega t) \qquad (2.16)$$

causes the voltage drop

$$v = \frac{1}{C} \int i \, dt = \frac{\widehat{I}}{\omega C} \sin(\omega t) = \widehat{V} \sin(\omega t) \quad \widehat{V} = \frac{\widehat{I}}{\omega C}; \quad V = \frac{I}{\omega C} \qquad (2.17)$$

The instantaneous power is an oscillation

$$p_{qc} = vi = \widehat{V}\widehat{I}\sin(\omega t)\cos(\omega t) = \frac{\widehat{V}\widehat{I}}{2}\sin(2\omega t) = Q_C\sin(2\omega t) \qquad (2.18)$$

with the amplitude

$$Q_C = VI = \frac{I^2}{\omega C} = \omega CV^2$$

The flow of energy into a capacitance is

$$w_{qc} = \int p_{qc}\,dt = \int Q_C\sin(2\omega t)\,dt = K' - \frac{Q_C}{2\omega}\cos(2\omega t) \qquad (2.19)$$

where K' is the integration constant. If at $\omega t = 0$, $v = 0$, then

$$w_{qc}|_{t=0} = \frac{Cv^2}{2} = 0$$

and

$$K' = w_{qc}|_{t=0} + \frac{Q_C}{2\omega}\cos(0) = \frac{Q_C}{2\omega}$$

thus

$$w_{qc} = \frac{Q_C}{2\omega}[1 - \cos(2\omega t)] \qquad (2.20)$$

We find that the mechanism of energy flow into the capacitance is similar to the flow of energy into an inductance. Under steady-state conditions, at the beginning of each voltage half-cycle, the capacitance starts to charge and maximum energy CV^2 is stored when the capacitor voltage reaches peak value. In the next quarter-cycle the capacitor is gradually discharged and is completely depleted at the next zero-voltage crossing.

The flow of energy in and out of the capacitance is shifted $T/4$ s with respect to the flow of energy in and out of the inductance. This means that when one connects in parallel a capacitance and an inductance that have equal reactive powers, i.e.

$$Q_C = \omega CV^2 = Q_L = \frac{V^2}{\omega L} = Q$$

the total energy "trapped" in the L–C system

$$w_q = w_{qc} + w_{qL} = \frac{Q}{2\omega}[1 - \cos(2\omega t) + 1 + \cos(2\omega t)] = \frac{Q}{\omega}$$

remains constant and $dw_q/dt = 0$. In this case no exchange of energy takes place between the L–C components and the source. This is exactly the case of parallel resonance, when

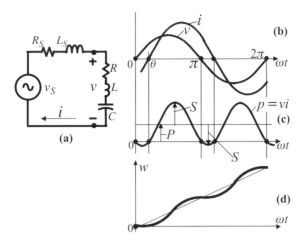

Figure 2.5 R–L–C Impedance supplied by a sinusoidal voltage: (a) Circuit. (b) Inductance voltage and current waveforms. (c) Instantaneous power waveform. (d) Energy.

$\omega L = 1/\omega C$. Once the source delivers to the L–C components the energy Q/ω, the oscillations of energy remain confined between L and C and, if the inductor and capacitor are lossless, no additional energy is required to support these oscillations. In this ideal situation the line current is nil.

2.4   The $R - L - C$ Loads

Next we will study the power and energy flowing into a series R–L–C branch supplied with the sinusoidal voltage (1), Fig. 2.5a. In this case the instantaneous current is

$$i = \widehat{I}\sin(\omega t - \theta) \tag{2.21}$$

where

$$\widehat{I} = \frac{\widehat{V}}{Z}; \quad Z = \sqrt{R^2 + X^2}; \quad X = \omega L - \frac{1}{\omega C}; \quad \tan\theta = \frac{X}{R}$$

The current i can be separated into two components: the active current i_p, in-phase with the voltage v, and the reactive current i_q, in-quadrature with the voltage v:

$$i = i_p + i_q$$

where

$$i_p = \widehat{I}_p sin(\omega t) = \widehat{I}\cos\theta\,\sin(\omega t); \quad \widehat{I}_p = \widehat{I}\cos\theta \tag{2.22}$$

is the active component and

$$i_q = -\widehat{I}_q \cos(\omega t) = -\widehat{I} \sin\theta \cos(\omega t); \quad \widehat{I}_q = \widehat{I} \sin\theta \qquad (2.23)$$

is the reactive component.

The instantaneous power also has two well-defined components

$$p = vi = v(i_p + i_q) = p_p + p_q \qquad (2.24)$$

where

$$p_p = vi_p = P[1 - \cos(2\omega t)] \qquad (2.25)$$

is the *instantaneous active power* having an expression identical to (2.6), that consists of active power and intrinsic power, and

$$p_q = vi_q = -Q \sin(2\omega t) \qquad (2.26)$$

is the *instantaneous reactive power* identical to (2.11).

The expressions of P and Q are

$$P = VI \cos\theta \quad \text{and} \quad Q = VI \sin\theta \qquad (2.27)$$

The active power P is the average value of p_p and is dissipated by R

$$P = RI^2 = R\frac{V}{Z}\frac{V}{Z} = V\frac{V}{Z}\frac{R}{Z} = VI \cos\theta \qquad (2.28)$$

The reactive power

$$Q = XI^2 = X\frac{V}{Z}\frac{V}{Z} = V\frac{V}{Z}\frac{X}{Z} = VI \sin\theta \qquad (2.29)$$

is the amplitude of p_q. Since p_q does not transfer energy that is converted by the load in other forms of energy, p_q can be categorized as a *nonactive instantaneous power*.

2.5 The Apparent Power

The waveforms of the voltage, current, and instantaneous power for a series R–L–C circuit are shown in Fig. 2.5b and c. The instantaneous power oscillates between the extremes $p_{min} = (P - S) \le 0$ and $p_{max} = (P + S) > 0$. The expression of the amplitude S of the instantaneous power oscillation becomes evident when we substitute (2.25) and (2.26) in (2.24) and obtain a detailed expression for the instantaneous power:

$$p = P - P \cos(2\omega t) - Q \sin(2\omega t) = P - S \cos(2\omega t - \theta) \qquad (2.30)$$

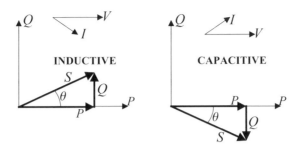

Figure 2.6 Power triangles: (a) Inductive load. (b) Capacitive load.

where

$$S = \sqrt{P^2 + Q^2} = VI; \quad \tan\theta = \frac{Q}{P} = \frac{X}{R} \tag{2.31}$$

The quantity S is called *apparent power*, has the unit VA (Volt-Ampere) and besides being equal to the amplitude of the instantaneous power oscillations (that is true only for sinusoidal conditions) it has a significant property related to the line power loss ΔP:

$$\Delta P = R_s I^2 = R_s \left(I_p^2 + I_q^2\right) = \frac{R_s}{V^2}(P^2 + Q^2) = \frac{R_s}{V^2}S^2 \tag{2.32}$$

meaning that the power loss in the line that supplies a load, or a cluster of loads that use a total apparent power S, is proportional to S^2. This obvious and simple property helps understand the true meaning of S as being the electrical quantity that relates the size of the load (kVA) with the size of the equipment needed to supply the required energy to the load. At the same time (2.32) reveals that both P and Q contribute to the energy lost in the line.

The three powers, P, Q and S, form a right-angle triangle, Fig. 2.6. If the current phasor lags the voltage (inductive load) $Q > 0$, (Fig. 2.6a). If the current leads the voltage (capacitive load) $Q < 0$, (Fig. 2.6b).

The power triangles representation leads to the concept of *complex power*[1]. This is a vectorial representation of the apparent power in the plane P–Q.

$$\mathbf{S} = P + jQ = S\angle\theta = VI\angle\theta$$

The phasors voltage, current and conjugate current phasors are

$$\mathbf{V} = V\angle 0°, \quad \mathbf{I} = I\angle{-\theta} \quad \text{and} \quad \mathbf{I}^* = I\angle\theta$$

---

[1] The complex power should not be confused with a phasor. A phasor usually represents a current or a voltage that has a perfectly sinusoidal time variation. The phasor is a segment with a length proportional with the amplitude or the rms value of the current or voltage. The phasor is like a wheel spoke that rotates with the angular velocity $\omega = 2\pi f$. In phasor diagrams the phasors are shown "frozen" at a certain moment in time, usually at $t = 0$.

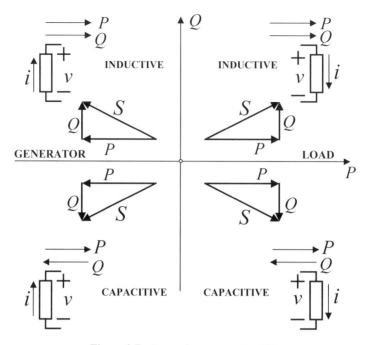

Figure 2.7   Power flow convention [2].

therefore

$$\mathbf{S} = \mathbf{V}\mathbf{I}^* = VI\angle\theta \quad \text{and}$$

$$P = \Re e\{\mathbf{V}\mathbf{I}^*\}; \quad Q = \Im m\{\mathbf{V}\mathbf{I}^*\} \tag{2.33}$$

Power-flow or load-flow studies compute at each bus or node of the modeled power system the complex power $\mathbf{S} = P + jQ$. These values show the direction of electric energy flow. The signs of P and Q are taken as summarized in Fig. 2.7 [2]. The convention for the direction of the active power flow is obvious: net energy is transferred in the positive direction of the flow. Active power exiting from a bus means that electric energy is generated or supplied from the bus to the network. The nonactive power p_q oscillates back and fro, its average value is nil, and in sinusoidal systems, Q represents the amplitude of the oscillations of p_q. Nevertheless, electrical engineers have attached a sign to Q: when the load is capacitive $Q < 0$ and when the load is inductive $Q > 0$. This convention led to the often mentioned, but physically incorrect, claim that "capacitors generate and inductors sink reactive power."

The expressions (2.32) and (2.33) show that if the reactive power Q is decreased while maintaining a constant active power P supplied to the load, then it is possible to minimize the supplying line power loss ΔP. Under such conditions if $Q \to 0$ the line power loss is reduced by $(R_s/V^2)Q^2$.

On the other hand, if one can keep both the line rms current I and the load rms voltage V constant and vary the phase angle θ, then the apparent power $S = VI$ remains unchanged. Under this constraint the power loss remains constant while the vectors P and Q slide along

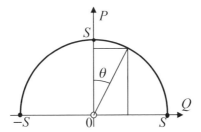

Figure 2.8 Active power vs. reactive power for a load with constant apparent power.

a circle $P^2 + Q^2 = S^2$, Fig. 2.8. If the reactive power is adjusted in the range $-S \leq Q \leq S$, the active power reaches its maximum at $Q = 0$ when $P = S$. All these observations lead to the following sound definition of S:

For single-phase systems, operating under sinusoidal conditions, the apparent power of a load or a cluster of loads supplied by a feeder is the maximum active power that can be transmitted through the feeder, while keeping the receiving end rms voltage and the feeder variable losses[2] constant. This definition can also be extended to a source: *The apparent power of a source is the maximum active power that can be supplied, or generated by, the source, while keeping its output voltage and the internal variable power losses constant.*

The above definition was introduced in a modified form by W. V. Lyon [3] in 1920, promoted by A. Liénard [4] in 1926, and later advocated by H. L. Curtis and F. B. Silsbee [5,6].

It cannot be emphasized enough that the apparent power is one of the most important electrical quantities that help the implementation of economic evaluations of transmission and conversion processes: electric energy billing and contracts are affected by the kVA demand. The cost of equipment is often expressed in \$/kVA. The compactness of a design is estimated in kVA/kg or kVA/m^3. Prototype and field testing, as well as standard compliance, pivot around the S–measurement.

From the beginnings of transformer and rotating machinery construction, the designers were aware of the significance of S and the importance of its accurate measurement. Arnold and later Richter [7] have developed key equations crucial for the process of shaping and seizing electrical equipment. Two famous design formulae come to mind:

For transformers;

$$A_{col} = C_T \sqrt{\frac{S}{f}} \tag{2.34}$$

where

A_{col} is the cross-sectional area of the transformer column (m^2),

f is the power system frequency (Hz),

---

[2] Feeder power loss includes the no-load losses, i.e. the iron-core losses of transformers and eventual dielectric losses. These losses are a function of V^2, hence independent of S. The variable losses are the losses proportional with I^2.

and for motors;

$$C_M D^2 L_i n = S \qquad (2.35)$$

where
   D is the diameter of the rotor (m),
   L_i is the equivalent length of the rotor (m),
   n is the rotor speed (rev/min) and

C_T, C_M are transformer and motor design constants, respectively. They are strongly affected by the heat transfer conditions and the physical properties of the materials involved in the construction of such equipment. Thus the efficiency, geometry, quality, and the amount of active materials and dielectrics, and ultimately the cost of electrical equipment, are functions of S and not of P.

A cardinal property of the apparent power that stems from (2.32) is the fact that *for a constant load voltage the total power loss in a feeder is a nearly linear function of S^2,* Fig. 2.9;

$$\Delta P_T \approx a + b S^2 \qquad (2.36)$$

where a and b are constants, (a represents the fixed losses). This simple approximation holds true for any types of power systems, any conditions, and any waveforms [8], and it should serve as a "go no-go gauge" for a simple, correct, and practical definition of S (see problem 2.15).

2.6   The Concept of Power Factor and Power Factor Correction

Electric utilities' goal is to deliver electric energy to their customers. When a transmission or a distribution line is built, capital is invested, and the investor, be it private or public, expects the investment to bear fruit. The energy lost in the conductors of feeders and transformers

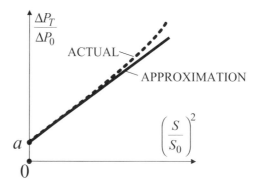

Figure 2.9 Total power loss in a feeder's conductors and transformers is nearly proportional to the apparent power squared.

translate in lost income; moreover, all these losses are converted into heat that, when excessive, causes premature aging of cable and winding insulation, or the annealing of the conductors; thus reducing the mechanical strength that is so critical for overhead applications. In overhead lines, increased conductors' temperature causes increased sags that may reduce the clearances to unacceptable levels. In cables, higher temperature means a more vulnerable dielectric to voltage surges and over voltages. Supplying lines are built to operate with a rated current-carrying capacity known as ampacity. When the end-users have loads characterized by large values of Q, the supplying line is not utilized correctly; the line may operate with a current near its rated ampacity without transferring much energy (kWh) to the end-users. From the economist's viewpoint this is a bad situation.

The figure of merit that can help to evaluate at a glance the utilization of a simple feeder or a transmission line is the ratio

$$\frac{W_p}{W_s} = \frac{\frac{1}{\tau}\int_0^\tau P\,dt}{\frac{1}{\tau}\int_0^\tau S\,dt} = \frac{\langle P\rangle}{\langle S\rangle} \le 1.0 \quad W_p \le W_s \tag{2.37}$$

Here W_p is the actual energy delivered to the observed load during the time τ and W_s is the hypothetical maximum possible energy that could be delivered through the line to its receiving-end during the same time τ, while keeping the energy lost in the supplying line unchanged and the receiving-end voltage unchanged. If the load voltage remains the same it means that the energy conversion process implemented by the load remains unchanged.

From (2.37) results the well known *power factor* definition:

$$PF = \frac{P}{S}$$

however, it should be understood that P and S rarely remain constant in time. These quantities fluctuate due to voltage and load variation, so for a specified period of time, the more correct expression of the power factor should use the mean value $\langle P\rangle$ and the equivalent value of S (see Chapter 7, Section 1.2, the Randomness Power).

An ac load supplied with the rms voltage V and characterized by the apparent power S and power factor PF can be represented by means of an equivalent resistance R in parallel with an equivalent reactance $X_L = \omega L$, Fig. 2.10a. Since

$$P = S(PF) = \frac{V^2}{R} \quad \text{and} \quad Q = S\sqrt{1-(PF)^2} = \frac{V^2}{X_L}$$

results

$$R = \frac{V^2}{P} = \frac{V^2}{S(PF)} \quad \text{and} \quad X_L = \frac{V^2}{Q} = \frac{V^2}{S\sqrt{1-(PF)^2}}$$

From the phasor diagram shown in Fig. 2.10b we see that the power factor

$$PF = \frac{P}{S} = \frac{VI_R}{VI} = \frac{I_R}{I} = \cos\theta \tag{2.38}$$

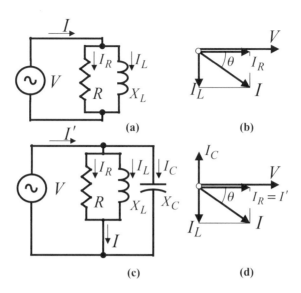

Figure 2.10 Equivalent R–L parallel load: (a) Circuit. (b) Phasor diagram for (a). (c) Power factor corrected load (compensated load). (d) Phasor diagram for (c).

This result is true only for circuits with sinusoidal waveforms.

Now let us assume that an amount of energy W is delivered to the load during the time $\tau = W/P$. Under this condition the line current is

$$I = \frac{S}{V} = \frac{\sqrt{P^2 + Q^2}}{V} = \sqrt{\left(\frac{V}{R}\right)^2 + \left(\frac{V}{X_L}\right)^2}$$

If a capacitance is connected in parallel with the load, Fig. 2.10c, and the capacitance is adjusted at resonance value $X_C = 1/\omega C = X_L = \omega L$, then the instantaneous currents i_C and i_L, that are $180°$ out of phase, cancel each other and the line current is minimized. The new rms line current is $I' = I_R = V/R < I$, Fig. 2.10d. This situation allows for the flow of more active power, i.e. another resistance, representing an additional load with unity power factor, can be connected in parallel such that the new total active power delivered to the receiving end of the line is increased from P to $P' = S$. During the time τ, the amount of energy supplied is $W' = P'\tau = S\tau$, and the energy W will be delivered in a shorter time $\tau' = (W/W')\tau$, Fig. 2.11a. It is easily observed that the power factor of the original (uncompensated load) is

$$PF = \frac{W}{W'} = \frac{\tau'}{\tau} = \frac{P}{P'} = \cos\theta \qquad (2.39)$$

Another simple, but meaningful, interpretation of the power factor, is to imagine the entire cross-sectional area A_S of the conductor proportional to the apparent power squared S^2, and the area $A_P < A_S$ proportional to the active power squared $P^2 < S^2$, Fig. 2.11b. One may assume all the active current crowded within the shaded sector A_P, i.e. only this portion of the cross-sectional area is utilized to transport energy. Evidently $PF = \sqrt{A_P/A_S}$. The entire cross-section of the conductor is utilized if the reactive power is eliminated and $P = S$.

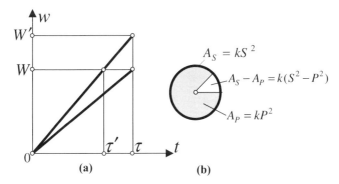

Figure 2.11 Geometrical interpretations of power factor: (a) Energy vs. time. (b) Conductor cross-section with area A_S. Shaded area A_P carries the active current. Blank area carries the nonactive current. The power factor $PF = \sqrt{A_P/A_S}$.

This geometrical interpretation leads to a simple *PF* mathematical expression: starting with the definition

$$PF = \frac{P}{S} = \frac{VI\cos(\theta)}{VI} = \sqrt{\frac{R_s[I\cos(\theta)]^2}{R_s I^2}} = \sqrt{\frac{\Delta P_c}{\Delta P}}$$

or

$$\frac{\Delta P_c}{\Delta P} = \frac{P^2}{S^2} = PF^2$$

where
   ΔP_c is the power loss after compensation,
   ΔP is the power loss before compensation.

Example 2.1   A single-phase source supplies 600 V, 60 Hz to a 200 kVA load with a power factor $PF = \cos\theta = 0.71$ lagging (i.e. inductive load). Determine the size of the capacitor connected in parallel with the load, that helps to raise the load power factor to $PF' = \cos\theta' = 0.90$.

The capacitor's reactive power Q_C is the amplitude of power oscillations 180° out of phase with the power oscillations Q of the load. From the phasor diagram and the triangles of powers, Fig. 2.12a, we find that

$$\tan\theta = \frac{Q}{S} \quad \text{and} \quad \tan\theta' = \frac{Q - Q_C}{S}$$

and from here results

$$Q_c = (\tan\theta - \tan\theta')S$$

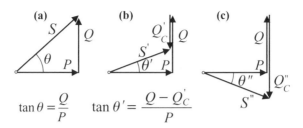

$$\tan\theta = \frac{Q}{P} \qquad \tan\theta' = \frac{Q - Q_C'}{P}$$

Figure 2.12 Power triangles: (a) Inductive load without compensation. (b) Inductive load with partial compensation. (c) Over compensated system.

For the given load we find

$$\tan\theta = \tan(\cos\ 0.71)^{-1} = 0.992; \quad \tan\theta' = \tan(\cos\ 0.9)^{-1} = 0.484$$
$$Q_C = (0.992 - 0.484)200 = 101.6\ \text{kvar}$$

The value of this capacitance is

$$C = \frac{Q_C}{\omega V^2} = \frac{101.6 \times 10^3}{376.99 \times 600^2} = 748.6\ \mu F$$

Unity power factor is obtained if $Q_C = Q = 198.4$ kvar. If $Q_C > Q$ the load will be over compensated, Fig. 2.12c and $PF < 1$, but leading.

2.7   Comments on Power Factor

Let us assume the general case where a linear load is supplied by a feeder with the impedance $R_s + jX_s$, Fig. 2.13a. We take as phase reference the load voltage phasor $\mathbf{V} = V\angle 0°$ leading the line current phasor $\mathbf{I} = I\angle -\theta$. The sending-end voltage phasor $\mathbf{V_s}$, Fig. 2.13b, is controlled by the electric utility and has the expression

$$\mathbf{V_s} = \mathbf{V} + (R_s + jX_s)(\cos\theta - j\sin\theta)I \qquad (2.40)$$

If a capacitance $C = I\sin\theta/\omega V = 1/(\omega^2 L)$ is connected in parallel with the load, Fig. 2.13c, the line current will be reduced and brought in-phase with the load voltage, Fig. 2.13d, and the load will operate at unity power factor. This improved condition makes "room" for an additional load: the new load, if also compensated to unity power factor, may be represented by a resistance R', Fig. 2.13c, that can be adjusted to yield a line current phasor $\mathbf{I} = I'\cos\theta + I_{R'} = I$, i.e. identical in magnitude to the initial line current, but in-phase with the load voltage. This means $I = V'/R_T$ where $R_T = RR'/(R + R')$. However, if the magnitude of the substation voltage phasor $\mathbf{V_s}$ remains unchanged, the new line current phasor $\mathbf{I} = I$ will modify the load voltage phasor to a new value:

$$V' = \sqrt{V_s^2 - (X_s I)^2} - R_s I$$

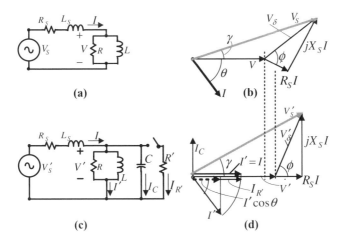

Figure 2.13 Equivalent R–L parallel load supplied by a line: (a) Circuit. (b) Phasor diagram. (c) Compensated circuit with additional load R'. (d) Phasor diagram for the compensated circuit.

If $\theta < 0$ then $V' > V$ and the load voltage remains unchanged only if the substation voltage is modified to a new value V_s'':

$$V_s'' = \sqrt{(V + R_s I)^2 + (X_s I)^2}$$ (2.41)

This means that the correct value of the additional load R' is found from the condition $R_T' = V/I$ with $R_T' = RR'/(R + R')$.

It is concluded from this analysis that in many practical situations the process of power factor correction is a two-step operation that requires the cooperation of both interested parties, consumer, and utility:

1. A capacitor or an active compensator device is installed in parallel with the load whose power factor must be corrected. This procedure—usually the end-user's responsibility—reduces the amplitude of the line current bringing its phasor in-phase with the load voltage phasor and at the same time causes the load voltage to increase by a certain increment.
2. The load voltage deviation from its initial value may be unacceptable to some consumers. The party responsible for the quality of the electric energy delivered has to maintain the customer's rms voltage within a permissible tolerance range (if the voltage is nonsinusoidal more constrains may apply). By means of substation transformer tap changers or voltage regulators, the voltage at the distribution system buses is maintained within a range that results in satisfactory operation of customer's loads and utilization equipment.

This conclusion sheds more light on the concept of the apparent power S, indicating that the apparent power is equal to an active power supplied to a load or a group of loads under ideal conditions (such loads may be considered fictive):

A. The load(s) power factor is compensated at $PF = 1$.
B. The rms line current must remain unchanged, hence the line power loss is kept unchanged. This condition implies that additional loads are connected in parallel with the observed load. The existing load compensated at $PF = 1$ takes a rms current smaller than the uncompensated load and to maintain the same line losses requires a load increase.
C. The load voltage is maintained unchanged. This requirement means that the supply voltage at the customer's mains has to be readjusted to a value that ensures equipment performance (electric energy conversion or generation) no different than the conversion prior to the power factor compensation.

One must be aware of the fact that the apparent power is not an actual physical quantity. In spite of its clear interpretation as a useful quantifier of conductor utilization, equipment sizing, geometry, and thermal stresses, and in spite of its impact on engineering economics, S is a mathematical definition, an expression that conveniently defines optimum conditions for the electric energy flow, or the evaluation of thermal stress in rotating machines, cables, and transformers. Modern instrumentation can measure any defined expression that is a function of actual electrical quantities, however, it is the responsibility of the engineering community to define and promote expressions that help provide an accurate overview of energy consumption, help increase energy savings, increase reliability of supply, enable improved energy management, and assist in the detection of fraud.

It is important to remember that historically, due to incandescent lamp dominance among loads, the constraint "load voltage unchanged" considered the rms voltage value, i.e. the thermoelectric effect. However, motors and rectifiers respond differently to voltage variations, distortions, and unbalance than a filament lamp resistance. The constraint "rms voltage to be constant" may be challenged and replaced with "load output unchanged," meaning that a motor will deliver, after the PF-compensation, the same torque and velocity, and a rectifier will deliver the same dc voltage. Probably the rigorous approach is to consider a voltage which facilitates the same active power conversion, i.e. the energy conversion process at the load remains unchanged. The output power is the same before and after the power factor correction.

Example 2.2 A single-phase load under normal operation is supplied with $\mathbf{V} = 440\angle 0°$ V, 60 Hz. The apparent power and the power factor of this load are $S = 44.0\,\text{kVA}$ and $PF = 0.707$ lagging ($P = 31.11$ kW and $Q = 31.11$ kvar). This load is connected to the substation via a dedicated line with an impedance $0.07 + j0.337\ \Omega$.

The load can be modeled by means of a resistance $R = V^2/P = 440^2/31110 = 6.223\ \Omega$ in parallel with an inductance $L = V^2/\omega Q = 440^2/377 \times 31110 = 16.509$ mH. The line current is $100\angle -45°$ A. The substation voltage (2.40) is

$$\mathbf{V_s} = 440 + (0.07 + j0.377)100\angle 45° = 472.107\angle 2.635°\ \text{V}$$

If a capacitance $C = 1/\omega^2 L = 426\ \mu F$ is connected in parallel with the load then unity power factor is obtained. In this case the line current is reduced to

$$\mathbf{I}' = \mathbf{V_s}/(0.07 + j0.377 + 6.223) = 74.886\angle 0.793°\ \text{A}$$

The 25.1% current reduction means 43.9% line power loss reduction. However, the load voltage increases to $V' = 6.223I' = 466.015$ V. This 5.91% load voltage jump is not necessarily "good news." Depending on the load type it may causes a 12.19% increase of the active power to 34.90 kW. Good engineering practice requires to restore the voltage magnitude within an acceptable range, closer to the rated value of 440 V, in which case the substation voltage has to be reduced to $V'_s = 445.95$ V. If a new load of $44.0 - 31.11 = 12.89$ kW, ($R' = 15\ \Omega$) is connected to help restore the current to its initial 100 A value, the substation voltage must be reduced to 448.27 V.

2.8   Other Means of Reactive Power Control and Compensation

In the past, synchronous machines were used to adjust the flow of the reactive energy w_q (2.14). The principle of operation can be explained with the help of Fig. 2.14 where an ideal synchronous motor—winding, core, and mechanical losses are negligible—is modeled using a synchronous EMF phasor $\mathbf{E} = E\angle - \delta$ in series with a synchronous reactance X_s. The motor is supplied with the voltage phasor $\mathbf{V_s} = V_s\angle 0°$. The current phasor is

$$\mathbf{I} = \frac{V_s - E\angle - \delta}{jX_s} = \frac{E}{X_s}\sin\delta - j\left(\frac{V_s - E\cos\delta}{X_s}\right)$$

where δ is the torque angle (Fig. 2.14).

The mechanical power developed by this ideal motor equals the active power

$$P = \Re e\left[\mathbf{V_sI^*}\right] = \frac{V_s E}{X_S}\sin\delta = V_s I\cos\theta \tag{2.42}$$

The instantaneous reactive power has the amplitude

$$Q = \Im m\left[\mathbf{V_sI^*}\right] = \frac{V_s^2 - V_s E\cos\delta}{X_S} = V_s I\sin\theta \tag{2.43}$$

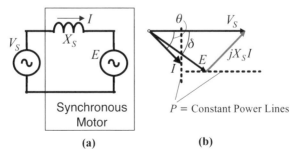

(a)   **(b)**

Figure 2.14   Ideal synchronous motor: (a) Per phase equivalent circuit. (b) Phasor diagram (line-to-neutral voltages).

The torque angle δ is a function both of E and the mechanical power P. If the motor is not loaded, $\delta = 0$, $\theta = \pm 90°$, $P = 0$ and

$$Q = V_s \frac{V_s - E}{X_s}$$

When the motor is overexcited, i.e. $E > V$, the reactive power $Q < 0$. In this case the synchronous motor operation is equivalent to a capacitance. The opposite is true for the underexcited case $E < V$, when $Q > 0$ and the motor operates like an inductance. The synchronous EMF E can be continuously adjusted by varying the field current. This important feature ensures continuous adjustment of Q. The main drawbacks of these "synchronous condensers" are the slow response time, bulkiness (low kg/kvar, low m³/kvar), and high annual investment cost ($/kvar).

Modern static compensators using solid state switching devices have replaced the "synchronous condensers." Their basic concept of reactive power flow control remains unchanged: the new technique is based on the fact that an adjustable alternating voltage synchronized to the network frequency can help to control the flow of the reactive instantaneous power. The controlled voltage is implemented using a power factor compensator that consists of a pulse-width-modulated solid state inverter or converter designed to control the rate of energy flow in or out of an energy storage component. The energy can be stored in a battery, a capacitor, a superconductive inductor, or a rotating mass. The system performs as an adjustable capacitor or inductor [9–13], the rate and direction of energy flow is controlled by adjusting the amplitude, and the phase of the voltage at the terminals of the power factor compensator. One of the major advantages of such a device is its fast response time. The major limitation is caused by the switching power loss in the solid-state devices (transistors or gate-turn-off thyristors).

The most general case is shown in Fig. 2.15. The uncompensated load $\mathbf{Z_L} = Z_L \angle \theta$ is supplied via a line with impedance $\mathbf{Z_s} = R_s + \jmath \omega L_s = Z_s \angle \phi$ by the voltage source $\mathbf{V_s}$. The phasor diagram, Fig. 2.15b, uses the load voltage phasor $\mathbf{V} = V \angle 0°$ as reference. The load

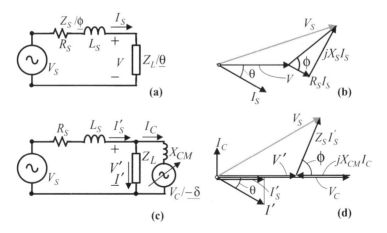

Figure 2.15 Reactive power compensated by means of static compensator with zero active power: (a) Uncompensated circuit. (b) Phasor diagram. (c) Circuit with compensator. (d) Phasor diagram for a perfectly compensated circuit.

voltage magnitude is a function of supply voltage phasor, line impedance, and the equivalent load impedance

$$\mathbf{V} = \frac{Z_L \angle \theta}{Z_L \angle \theta + Z_s \angle \phi} \mathbf{V_s}$$

The rms load voltage can be written in the form

$$V = V_s / F(Z_s, Z_L)$$

where

$$F(Z_s, Z_L) = \sqrt{1 + \left(\frac{Z_s}{Z_L}\right)^2 + \frac{2Z_s}{Z_L} \cos(\phi - \theta)}$$

The active and reactive load powers are

$$P = V I_s \cos\theta = \left(\frac{V_s}{F(Z_s, Z_L)}\right)^2 \frac{\cos\theta}{Z_L}$$

$$Q = V I_s \sin\theta = \left(\frac{V_s}{F(Z_s, Z_L)}\right)^2 \frac{\sin\theta}{Z_L}$$

(2.44)

In the next step a static compensator with a variable voltage phasor $\mathbf{V_c} = V_c \angle - \delta$ and a series equivalent reactance X_{cM} is connected in parallel with the load Z_L, Fig. 2.15c. The phasor $\mathbf{V_c}$ is brought in-phase with the load voltage phasor $\mathbf{V'}$ (i.e. $\delta = 0$) and adjusted to inject the current

$$I_C = (V_c - V') / X_{cM} = I' \sin\theta$$

The resulting line current is reduced to

$$I'_s = I' \cos\theta \quad \text{with} \quad I' = V' / Z_L$$

In this manner the unity power factor is obtained, but the voltage phasors (Fig. 2.15d) conform now to equation

$$V_s^2 = (V' + Z_s I'_s \cos\phi)^2 + (Z_s I'_s \sin\phi)^2$$

that leads to the new load voltage

$$V' = V_s / F'(Z_s, Z_L)$$

where

$$F'(Z_s, Z_L) = \sqrt{1 + \left(\frac{Z_s}{Z_L} \cos\theta\right)^2 + \frac{2Z_s}{Z_L} \cos\theta \cos\phi}$$

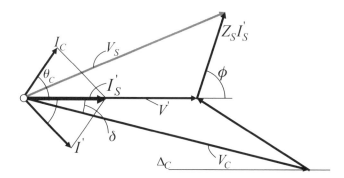

Figure 2.16 Power conditioner (converter able to compensate nonactive power and to deliver active power): Phasor diagram.

Since $F(Z_s, Z_L) > F'(Z_s, Z_L)$ results that $V' > V$ and

$$P' = \left(\frac{V_s}{F'(Z_s, Z_L)} \right)^2 \frac{\cos \theta}{Z_L} > P$$

and the supplied reactive power $Q' = 0$.

Note: Some power electronics devices have the capability to vary the phase angle δ and the amplitude of the phasor $\mathbf{V_c}$, Fig. 2.16. Such modern converters and power conditioners [12] are designed to control both active and nonactive power flow. Photovoltaic generators, fuel cells, and certain converters used for adjustable speed drives have the ability to generate or to convert electric energy while providing the desired nonactive power needed to maintain the required voltage profile along a feeder, to maintain voltage stability, or to correct the *PF* at the "Point of Common Coupling."

The phasor diagram for such systems is shown in Fig. 2.16. The converter operates with the voltage $\mathbf{V_c} = V_c\angle - \delta$, $\delta \neq 0$ and injects the current $\mathbf{I_c} = (\mathbf{V'} - \mathbf{V_c})/jX_{CM}$. By sliding the phasor $\mathbf{V_c}$ along the power line Δ_C the converter's active power is kept constant and the reactive component of the current, $I_{cq} = I_c \sin(\theta_c) = I' \sin(\theta)$, in which case the total current I'_s is in-phase with the load voltage V'.

2.9 Series Compensation

Power factor compensation by means of a series connected capacitor is possible in situations where the load voltage is allowed to vary in function of current. If the equivalent load is reduced to a series R–L–C circuit supplied by the rms voltage V_s, the needed series capacitance produces series resonance and has the value $C_\sigma = 1/\omega^2 L$. The rms current is $I = V_s/R$ and the capacitance and inductance voltage are $V_C = V_L = (\omega L/R)V_s$. Some power electronics circuits use this connection. The major drawback is observed when $\omega L \gg R$ and this approach may not be feasible since $V_C \gg V_s$.

If the equivalent load is a parallel R–L circuit (some induction furnaces designs fit well this case) the load impedance is

$$\mathbf{Z} = \frac{\omega^2 L^2 R + j\omega L R^2}{R^2 + (\omega L)^2}$$

and the required capacitance is

$$C_\rho = \frac{R^2 + (\omega L)^2}{\omega^2 L R^2}$$

The input current phasor is in-phase with the voltage and its rms magnitude is

$$I = [1 + (R/\omega L)^2] V_s / R$$

causing the capacitance and load voltages

$$V_C = \frac{R}{\omega L} V_s; \quad V_L = \sqrt{1 + (R/\omega L)^2}\, V_s$$

More can be learned about different methods of reactive power control from references [10–14].

2.10 Reactive Power Caused by Mechanical Components that Store Energy

Let us assume the hypothetical electromechanical system sketched in Fig. 2.17a where the frequency of the sinusoidal voltage v is low enough to allow the rotor of a dc motor to rotate back and fro. This circuit is governed by two equations: First, Kirchhoff's voltage law for the electric circuit

$$v = \widehat{V} \sin(\omega t) = Ri + e = Ri + K I_F \Omega \qquad (2.45)$$

and the second is the torques balance

$$\Upsilon = K I_F i = J \frac{d\Omega}{dt} \qquad (2.46)$$

where
e is the motor's back EMF,
K is the motor constant (Vs/A rad) or (Nm/A^2),
I_F is the field current (A),
Ω is the angular velocity of the motor (rad/s),
Υ is the torque developed by the motor, (Nm), and
J is the momentum of inertia (kgm^2).

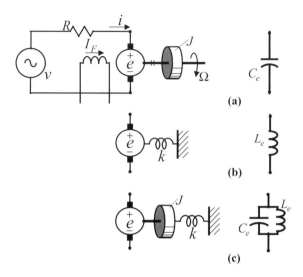

Figure 2.17 Electromechanical loads and their equivalent circuit: (a) Flywheel—equivalent capacitance. (b) Torsion bar—equivalent inductance. (c) Flywheel and torsion bar—equivalent parallel L_e–C_e. Source: A. E. Emanuel, "Powers in Nonsinusoidal Situations: A Review of Definitions and Physical Meaning," IEEE Trans. On Power Delivery, Vol.5, No.3, July 1990. Copyright 1990, IEEE.

From (2.46) results that the angular velocity Ω and the current i are correlated,

$$\Omega = \frac{K I_F}{J} \int i \, dt \tag{2.47}$$

Substitution of (2.47) in (2.45) gives

$$\widehat{V} \sin(\omega t) = Ri + \frac{(K I_F)^2}{J} \int i \, dt = Ri + \frac{1}{C_e} \int i \, dt \tag{2.48}$$

We observe that such an electromechanical system is equivalent to a capacitor with a capacitance

$$C_e = \frac{J}{(K I_F)^2} \quad (F)$$

The instantaneous current equation is

$$i = \frac{\widehat{V}}{Z} \sin(\omega t + \theta); \quad Z = \sqrt{R^2 + (1/\omega C_e)^2}; \quad \tan \theta = R C_e \omega$$

and the instantaneous reactive power (see (2.26) and (2.27)) is

$$p_q = (V I \sin \theta) \sin(2\omega t) = Q \sin(2\omega t) \tag{2.49}$$

where

$$Q = \frac{V^2}{Z^2} \frac{1}{C_e \omega} = \frac{(VKI_F)^2}{J\omega Z^2}$$

The mechanical torque (2.46)

$$\Upsilon = KI_F i = KI_F \frac{\widehat{V}}{Z} \sin(\omega t + \theta) = J\frac{d\Omega}{dt}$$

yields

$$\Omega = KI_F \frac{\widehat{V}}{Z} \int \sin(\omega t + \theta) = \frac{-KI_F \widehat{V}}{ZJ\omega} \cos(\omega t + \theta)$$

and the instantaneous mechanical power produced by the motor is

$$p_m = \Upsilon \Omega = \frac{-(KI_F V)^2}{J\omega Z^2} \sin(2\omega t + 2\theta) = -Q\sin(2\omega t) \qquad (2.50)$$

Comparing the power p_m with the instantaneous reactive power p_q we observe that, while 180° out of phase, the rate of flow of the observed energies has identical amplitudes Q.(See problem 2.7).

If the flywheel is replaced with a torsion bar with the spring constant k (Nm/rad), Fig. 2.17b, the torque equation (2.46) becomes

$$\Upsilon = KI_F i = k \int \Omega \, dt$$

hence

$$\Omega = \frac{1}{k}\frac{d\Upsilon}{dt} = \frac{KI_F}{k}\frac{di}{dt}$$

Kirchhoff's voltage law gives

$$\widehat{V}\sin(\omega t) = Ri + KI_F\Omega = Ri + \frac{(KI_F)^2}{k}\frac{di}{dt}$$

We observe that this time the electromechanical system can be represented by an equivalent inductance with an inductance,

$$L_e = \frac{(KI_F)^2}{k} \quad (H)$$

Next let us couple the motor shaft with a flywheel and a torsion bar, Fig. 2.17c. The new torque equation is

$$\Upsilon = J\frac{d\Omega}{dt} + k\int \Omega \, dt \qquad (2.51)$$

This time we will make use of Laplace transform. Equation (2.51) becomes

$$\Upsilon(s) = KI_F I(s) = J\Omega(s)s + \frac{k}{s}\Omega(s)$$

or

$$\Omega(s) = \frac{KI_F I(s)}{Js + \dfrac{k}{s}}$$

that substituted in (2.45) gives

$$V(s) = RI(s) + (KI_F)^2 \frac{I(s)}{Js + \dfrac{k}{s}}$$

The second term

$$(KI_F)^2 \frac{I(s)}{Js + \dfrac{k}{s}} = \frac{\dfrac{(KI_F)^2}{k}s \, \dfrac{(KI_F)^2}{Js}}{\dfrac{(KI_F)^2}{k}s + \dfrac{(KI_F)^2}{Js}} I(s) = \frac{L_e s \dfrac{1}{C_e s}}{L_e s + \dfrac{1}{C_e s}} I(s)$$

proves that the dc motor–torsion bar–flywheel can be correctly modeled with an equivalent inductance in parallel with an equivalent capacitance. The resonance angular frequency is

$$\omega_0 = \frac{1}{\sqrt{L_e C_e}} = \sqrt{\frac{k}{J}}$$

and equals the mechanical resonance frequency.

The results obtained in this section can be extended to other similar situations; whenever the electric energy flow reverses direction every quarter-cycle, i.e when the electric energy is stored via reversible conditions in any form of energy (electromagnetic, kinetic, potential, thermal, etc.), such oscillations of energy are always supported by instantaneous reactive power.

2.11 Physical Interpretation of Instantaneous Powers by Means of Poynting Vector

It was explained in Chapter 1 that the electric energy is transmitted through the dielectric that separates and surrounds the conductors (air, polymers, ceramics, mineral oil, etc.). The flux of electromagnetic energy propagates along the conductors and the density of its rate of flow (i.e. the power density) at any given point in space is quantified by a vector $\vec{\wp} = \vec{E} \times \vec{H}$ measured in W/m^2. The conductors' role is to produce the electric and magnetic fields, E and H. In a simplistic manner one may consider the conductors as a "rail-road" or a wave-guide meant to facilitate the propagation of energy from source to load.

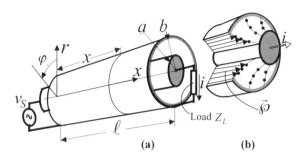

Figure 2.18   Coaxial cable supplying an ac load: (a) Three-dimensional view. (b) Poynting vectors inside the cable's dielectric.

In single-phase circuits with sinusoidal waveforms the power flow is characterized by the quantities P, Q and S. In this section we will try to determine the correlation between the electromagnetic energy wave—characterized by the Poynting vector distribution—and the powers P, Q and S.

We return now to the coaxial cable, Fig. 2.18a, that connects an impedance $Z_L\angle\theta$ with a sinusoidal voltage source $v_s = \widehat{V}\sin(\omega t)$ providing the line current

$$i = \widehat{I}\sin(\omega t - \theta); \quad \widehat{I} = \widehat{V}/Z_L \tag{2.52}$$

If the coaxial cable length ℓ is significantly shorter than the wavelength $\lambda \approx 1/(f\sqrt{\mu\epsilon})$, (where $f = \omega/2\pi$ Hz and μ and ϵ are the magnetic permeability and the dielectric permitivity of the insulating material that separates the conductors), and if the cable's resistance, inductance, and capacitance can be neglected, then the electric and magnetic field distributions along the cable are functions of voltage v, current i, and radius r:

$$\vec{E} = \frac{K_E}{r}v_s\vec{1}_r; \quad K_E = \frac{1}{\ln(b/a)} \tag{2.53}$$

and

$$\vec{H} = \frac{K_H}{r}i\,\vec{1}_\varphi; \quad K_H = \frac{1}{2\pi} \tag{2.54}$$

Since $\vec{E}$ and $\vec{H}$ are perpendicular, the Pointing vector within the dielectric can be readily obtained

$$\vec{\wp} = \vec{E} \times \vec{H} = \frac{K_E K_H}{r^2}v_s i\,\vec{1}_x = \frac{K_E K_H}{r^2}\widehat{V}\widehat{I}\sin(\omega t)\sin(\omega t - \theta)\vec{1}_x$$

$$= \frac{K_E K_H}{r^2}[P - P\cos(2\omega t) - Q\sin(2\omega t)]\vec{1}_x \tag{2.55}$$

with P and Q defined in (2.27). From (2.55) is learned that at any point in the dielectric volume $\pi(b^2 - a^2)\ell$, the Poynting vector is oriented parallel to the conductors and its magnitude is inversely proportional to the radius squared, Fig. 2.18b; moreover, we recognize that $\vec{\wp}$ has an active and a reactive component proportional to the instantaneous powers p_p and p_q, respectively.

$$\vec{\wp}_p = \frac{K_E K_H}{r^2} \, p_p \, \vec{1}_x = \frac{K_E K_H}{r^2} \, P[1 - \cos(2\omega t)] \, \vec{1}_x \tag{2.56}$$

and

$$\vec{\wp}_q = \frac{K_E K_H}{r^2} \, p_q \, \vec{1}_x = \frac{-K_E K_H}{r^2} \, Q \sin(2\omega t) \, \vec{1}_x \tag{2.57}$$

The flux of the Poynting vector through the cross-sectional area $\pi(b^2 - a^2)$ provides the total instantaneous active power p supplied to the load

$$-\int_a^b \vec{\wp} \, (2\pi r \, dr)(-\vec{1}_x) = 2\pi \, K_E K_H v_s i \int_a^b \frac{r \, dr}{r^2} = \frac{2\pi V_s i}{2\pi \ln(b/a)} \ln(b/a) = v_s i = p$$

If one calculates the flux carried by the active component, $\vec{\wp}_p$ will find, as expected, that the active component $\vec{\wp}_p$ "impinges" the active power to the load,

$$-\int_a^b \vec{\wp}_p \, (2\pi r \, dr)(-\vec{1}_x) = P[1 - \cos(2\omega t)]$$

and the time average of the Poynting vector is proportional to the active power P,

$$\langle \vec{\wp} \rangle = \langle \vec{\wp}_p \rangle = \frac{K_E K_H}{r^2} P \vec{1}_x$$

The active Poynting vector (2.56) pulsates with a double frequency 2ω and its electromagnetic wave always moves unidirectionally toward the load, i.e. $\vec{\wp}_p \geq 0$. A three-dimensional representation of the propagation of $\vec{\wp}_p$ in time is shown in Fig. 2.19a.

In the same manner we determine the flux carried by the reactive component, $\vec{\wp}_q$

$$-\int_a^b \vec{\wp}_q \, (2\pi r \, dr)(-\vec{1}_x) = -Q \sin(2\omega t)$$

This reactive component of the electromagnetic wave pulsates also at double frequency, but moves back and forth between the load and the source—a quarter of cycle from the load and the next quarter of cycle toward the load, Fig. 2.19b. This exchange or oscillations of energy between source and load take place on the account of the energy stored in and returned from inertive components such as L, C or electromechanical equipment. Over each half-cycle the net energy transferred by the reactive Poynting vector $\vec{\wp}_q$ is nil.

The electromagnetic wave inside the cable propagates with an astounding velocity

$$u \approx \frac{1}{\sqrt{\mu \epsilon}} = \frac{c}{\sqrt{\mu_r \epsilon_r}}$$

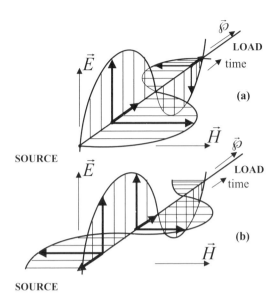

Figure 2.19 Time-variation of the field vectors $\vec{E}$, and $\vec{H}$: (a) Load with $PF = 1$. (b) Load with $PF = 0$.

where $c \approx 3 \times 10^8$ m/s is the velocity of light, $\mu_r = \mu/\mu_0$, $\epsilon_r = \epsilon/\epsilon_0$ and $\mu_0 = 4\pi\,10^{-7}$ H/m, $\epsilon_0 = 8.85 \times 10^{-12}$ F/m. For example, if $\mu_r = 1$ and $\epsilon_r = 2$ we obtain $u \approx 2.1 \times 10^8$ m/s.

Now we shall consider the more realistic case when the cable's impedance $\mathbf{Z_s} = R_s + J X_s = Z_s \angle \phi$, $\tan \phi = X_s/R_s$ is included in our model. This means that the load voltage is affected by the voltage drop v_δ across the cable; moreover, the voltage amplitude along the cable is not constant. At a point located x meters from the sending end of the cable the phasor voltage $\mathbf{V_x}$ (see Figs. 2.20 and 2.13) is

$$\mathbf{V_x} = \mathbf{V_s} - \frac{x}{\ell}\mathbf{V_\delta} \quad \text{where} \quad \mathbf{V_\delta} = \mathbf{Z_s I}; \tag{2.58}$$

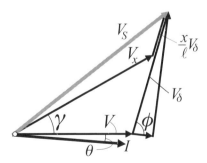

Figure 2.20 Phasor diagram for an inductive load supplied with voltage and current phasors $\mathbf{V}$ and $\mathbf{I}$ from a voltage source phasor $\mathbf{V_S}$ via a line with length ℓ and impedance $\mathbf{Z}_S = R_S + J X_S$. The phasor voltage at a distance x from the sending end is $\mathbf{V}_x$.

and the instantaneous voltage values are

$$v_\delta = Z_s \hat{I} \sin(\omega t - \theta + \phi) \tag{2.59}$$

$$v_x = \hat{V}_s \sin(\omega t + \gamma) - \frac{x}{\ell} v_\delta \tag{2.60}$$

The electric field within the dielectric of this lossy cable is now dependent on the voltage v_x, radius r and distance x. With the assumption that the cable impedance Z_s is equally shared by the inner and the outer conductor, one finds two components for the electric field [15,16] (see Appendix I)

$$\vec{E} = \vec{E}_r \vec{1}_r + \vec{E}_x \vec{1}_x$$

where

$$E_r = \frac{K_E}{r} v_x \quad \text{and} \quad E_x = F(r) v_\delta$$

are the radial and the axial component, respectively and

$$K_E = \frac{1}{\ln(b/a)}; \quad F(r) = \frac{\ln(ab/r^2)}{2\ell \ln(b/a)}$$

is a function of the radius r for $a \leq r \leq b$.

The magnetic field distribution is given by (2.54). The two components of $\vec{E}$ lead to two components for the Poynting vector

$$\vec{\wp} = \vec{E} \times \vec{H} = \vec{E}_r \times \vec{H} + \vec{E}_x \times \vec{H} = \vec{\wp}_x + \vec{\wp}_r$$

an axial component

$$\vec{\wp}_x = \vec{E}_r \times \vec{H} = \frac{K_E K_H}{r^2} v_x i \, \vec{1}_x \quad \vec{1}_r \times \vec{1}_\varphi = \vec{1}_x \tag{2.61}$$

and a radial component

$$\vec{\wp}_r = \vec{E}_x \times \vec{H} = \frac{-K_H}{2\ell \ln(b/a)} \frac{1}{r} \ln\left(\frac{ab}{r^2}\right) v_\delta i \, \vec{1}_r \quad \vec{1}_x \times \vec{1}_\varphi = -\vec{1}_r \tag{2.62}$$

Since v_x varies along the cable so does $\wp_x$. When we deal with a lossy feeder supplying an inductive load, we find that the intensity of the Poynting vector is slightly diminishing as it advances along the cable from $x = 0$ to $x = \ell$. The rate of the drop in intensity follows exactly the rate of the voltage drop along the cable.

The radial component $\wp_r$, is not a function of x, nevertheless it plays a very important role: it transfers energy to conductors and covers the line losses ΔP. We observe from (2.62) that on the cylindrical surface with $r_0 = \sqrt{ab}$ (where $\ln(ab/r_0^2) = 0$) the radial component $\wp_r = 0$. The PV flux lines on the surface of this cylindrical envelope with radius r_0 are perfectly parallel

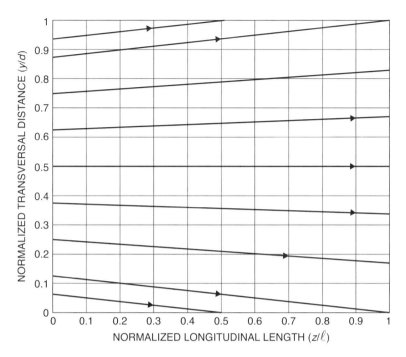

Figure 2.21 Poynting vector stream lines for two flat parallel conductors [16]. $v_\delta/v_s = 0.25$ and the space between the conductors is 0.20ℓ. Source: A. E. Emanuel, "Poynting Vector and the Physical Meaning of Nonactive Powers," IEEE Trans. On Instrumentation and Measurement, Vol. 54, No.4, August 2005, pp. 1457–62. Copyright 2005, IEEE.

to the conductors. However, for $r < \sqrt{ab}$ the radial Poynting vector is oriented toward the inner conductor, $\wp_r < 0$, and for $r > \sqrt{ab}$ the radial Poynting vector is oriented toward the outer conductor, $\wp_r > 0$. This observation gives a clear picture: energy flux lines inrush from the source-end through the dielectric of the cable and are slightly bending toward the inner and the outer conductors. The flux lines that end touching the conductors transfer inside the conductors the energy that covers Joule and eddy-current losses, as well as the energy stored in and returned from the electromagnetic field located within the conductors (Fig. 2.21).

Substitution of (2.52) and (2.59) in (2.62) gives the following expression for the radial component:

$$\wp_r = \frac{-K_H F(r)}{r} v_\delta i = \frac{-K_H F(r)}{r} Z_s \hat{I}^2 \sin(\omega t - \theta) \sin(\omega t - \theta + \phi)$$

$$= \frac{-K_H F(r)}{r} Z_s I^2 [\cos\phi - \cos(2\omega t - 2\theta + \phi)]$$

and since $Z_s \cos\phi = R_s$ and $Z_s \sin\phi = X_s$ results

$$\wp_r = \frac{-K_H F(r)}{r} \left\{ [R_s I^2 - Z_s I^2 [\cos\phi \cos(2\omega t - 2\theta) - \sin\phi \sin(2\omega t - 2\theta)] \right\}$$

substituting $I^2 = (P^2 + Q^2)/V^2$ yields

$$\wp_r = \frac{-K_H F(r)}{r} \left\{ [R_s \frac{P^2 + Q^2}{V^2} [1 - \cos(2\omega t - 2\theta)] + X_s \frac{P^2 + Q^2}{V^2} \sin(2\omega t - 2\theta) \right\} \quad (2.63)$$

The last equation is a significant result for our analysis. It shows that the radial Poynting vector has two major components. The first carries the line power losses $R_s I^2 = R_s S^2/V^2$ and the intrinsic power tied to it. The second component has no mean value, it does not contribute to the cable power loss, and consists of the power oscillations due to the line inductance $L_s = X_s/\omega$.

It can be easily proved that the radial component covers the power loss $R_S I^2$. The flux of the Poynting vector impinging on the inner conductor surface $2\pi \ell a$ is:

$$\Delta p_a = -\vec{\wp}_r|_{r=a} \, 2\pi \ell a \vec{1}_r = \frac{2\pi \ell a \ln(ab/a^2)}{2\pi \, 2\ell a \ln(b/a)} F(S, V, \theta, \omega t) = \frac{1}{2} F(S, V, \theta, \omega t) \quad (2.64)$$

where

$$F(S, V, \theta, \omega t) = \{ [R_s[1 - \cos(2\omega t - 2\theta)] + X_s \sin(2\omega t - 2\theta) \} \frac{S^2}{V^2}$$

and the flux of the Poynting vector impinging on the outer conductor surface $2\pi \ell b$ is:

$$\Delta p_b = -\vec{\wp}_r|_{r=b} \, 2\pi \ell b(-\vec{1}_r) = \frac{-2\pi \ell b \ln(ab/b^2)}{2\pi \, 2\ell b \ln(b/a)} F(S, V, \theta, \omega t) = \frac{1}{2} F(S, V, \theta, \omega t)$$

$$(2.65)$$

The total instantaneous power impinged on the two conductors is

$$\Delta p = \Delta p_a + \Delta p_b = F(S, V, \theta, \omega t)$$

with the mean value $\Delta P = R_s S^2/V^2$.

From (2.63) we also learn that $\wp_r$ can be separated into two components, one connected with the active power P and the other connected with the reactive power Q,

$$\wp_r = \wp_{rP} + \wp_{rQ}$$

The first radial component has the expression

$$\wp_{rP} = \frac{-K_H F(r)}{r} \left\{ [R_s \frac{P^2}{V^2} [1 - \cos(2\omega t - 2\theta)] + X_s \frac{P^2}{V^2} \sin(2\omega t - 2\theta) \right\}$$

and is the result of the interaction between the axial electric field E_x and the magnetic field produced by the active component of the current, i.e. it is a function of P.

The second radial component

$$\wp_{rQ} = \frac{-K_H F(r)}{r} \left\{ [R_s \frac{Q^2}{V^2}[1 - \cos(2\omega t - 2\theta)] + X_s \frac{Q^2}{V^2} \sin(2\omega t - 2\theta) \right\}$$

is the result of the interaction between the axial electric field and the magnetic field produced by the reactive component of the current, hence a function of Q.

We notice that both radial components supply the line conductors with active and reactive power. This observation has its physical interpretation: if we assume a pure inductive load, $P = 0$, $2\theta = 180°$ results that $\wp_{rP} = 0$, however, the component $\wp_{rQ}$ covers the line power loss $\Delta P = (Q^2/V^2)R_s$ as well as the oscillations of energy between the source and the inductance L_s.

In Fig. 2.22 the paths of instantaneous powers carried by the electromagnetic waves flowing through the dielectric that surrounds the line conductors are sketched for the general case of $R - L$ load. The solid lines mark the unidirectional flow of the waves that carry net energy from the source to the load and conductors. The dashed lines report on power that oscillates to and fro between load inductance L and source or line inductance X_s/ω and source. The line resistance R_s receives energy from the radial components $\wp_{rp}$ and $\wp_{rq}$. The same holds true for the reactive power exchanged with X_s/ω.

In Fig. 2.23 the flow of energy in a line supplying a compensated load is sketched, where $\wp_q = 0$. The load reactive power component is confined to the space surrounding the load and the power factor compensation device. The reactive instantaneous power exchanged with the line inductance X_s/ω is tied with the axial component of the Poynting vector.

Finally Fig. 2.24 shows a series capacitance compensated load at $PF = 1$. The load reactive power is confined to the space near the load and capacitor.

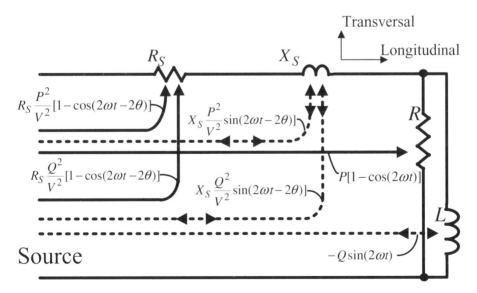

Figure 2.22 Electromagnetic energy flow (Poynting vector flow). Solid line: Unidirectional flow (active power). Dashed line: Oscillations (nonactive power). General case: uncompensated parallel R–L load.

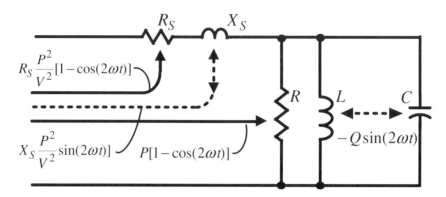

Figure 2.23 Electromagnetic energy flow (Poynting vector flow). Compensated load at $PF = 1$.

To understand the next chapters, which discuss power components and their meaning, it is important to realize that for single-phase sinusoidal conditions both components, P and Q, quantify not only the longitudinal flow of W/m^2, but also the transversal flow (in and out the line conductors). Since the Poynting vector describes electromagnetic radiation—the density of the rate of flow of electromagnetic energy—and since this vector has a definite physical meaning, so do all the components of this vector.

One last comment: The PV theory should not be viewed as the necessary, or the recommended, concept on which electric instruments designs are based (though some noninvasive measurement methods do rely on PV measurements). The spatial distribution of PV and its time variation have been explained in this section with one thought in mind: the in-depth understanding of the very physical mechanism that governs the energy flow, this helping further to discern among the different components of power.

The following comparison is made to emphasize the significance of Poynting vector: Assume a resistor with a certain three-dimensional geometry. The resistor has a resistance R and is supplied with the voltage V, in which case the current is $I = V/R$. This result is correct,

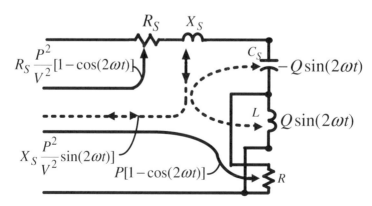

Figure 2.24 Electromagnetic energy flow (Poynting vector flow). Series capacitance compensated load at $PF = 1$.

but gives information limited to the macroscopic performance of the observed system. If one wants to determine spots of high current density, or to map the detailed streamlines flow of current, it will be necessary to remember the basic expression of the current density vector: $\vec{j} = \vec{E}/\rho$.

The Poynting vector $\vec{\wp} = \vec{E} \times \vec{H}$ is to "Power Theory," $P = RI^2 = VI = V^2/R$ what Maxwell's equations are to Ohm, Kirchhoff, and Faraday' laws.

2.12 Problems

2.1 Compute the equivalent R–L parallel circuit of a single-phase load, 240 V, 60 Hz, 5 kW, $PF = 0.60$, and determine the value of the parallel connected capacitance that will help increase the power factor to $PF = 1.0$ and $PF = 0.9$.

2.2 A load $\mathbf{Z} = (1 + j)\ \Omega$ is supplied by a 240 V, 60 Hz voltage source. The connecting line's impedance is $\mathbf{Z_s} = (0.01 + j0.06)\ \Omega$. Compute line current I_s, load voltage V, the load powers S, P, Q and the PF as well as the line power loss, ΔP. Use a capacitor to correct the power factor to $PF' = 1.0$. Repeat the computations. Compare the two sets of results. Next connect an additional resistance in parallel such that the rms line current equals the initial rms current. Repeat computations.

2.3 A single-phase motor operates at 120 V, 60 Hz, with 70% efficiency and power factor $PF = \cos\theta = 0.75$ lagging. The power delivered to mechanical load is 0.40 HP.

- Sketch the oscillograms $v = v(t)$ and $i = i(t)$. Compute and sketch the time-variations of the instantaneous powers p_p, p_q and p.
- Compute the value of the capacitance C that improves the power factor to $PF' = \cos\theta' = 0.95$.
- Compute and sketch the new instantaneous powers p'_p, p'_q and p'.

2.4 A nine-mile long, 7.9 kV single-phase feeder supplies a large number of customers assumed to be uniformly distributed along the feeder. The specific load is $\mathbf{S} = (80 + j60)$ kVA/mi. Determine the best location (measured in mi from the substation) and the best size of the capacitance that will help minimize the power loss in the feeder.

2.5 A small industrial facility operates with $P = 60$ to 140 kW and $Q = 70$ to 90 kvar. A photovoltaic generator which also has the capability to produce reactive power will be installed to reduce the input active power and to improve the PF. The photovoltaic generator operates in the second and third quadrant, Fig. 2.7, with $P = 0$ to 50 kW and $Q = 0$ to ± 50 kvar. Determine the possible range of PF values.

2.6 Write the general expression for the instantaneous power $p = vi$ delivered to a series R–C circuit. Next compute the instantaneous power $p_R = v_R i$ and $p_C = v_C i$. You will notice that p_R and p_p have equal average values and equal amplitudes, but they are not in-phase. Similar observations are made for p_q and p_C. Prove that

$$p = vi = p_p + p_q = v_R i + v_C i = P - S\cos(2\omega t + \theta)$$

2.7 A typical shaking table is an electromechanical system that consists of a cylindrical coil and a concentric cylindrical permanent magnet mechanically connected with the table. The magnet and the table have a total mass m (kg). The motion of the permanent magnet is opposed by a linear spring with a spring constant k (N/m). When the solenoid is energized the permanent magnet is attracted or repelled by the solenoid. The force on the magnet, $F \approx Ki$, is oriented along the axis of symmetry. The coil has a resistance R and an inductance L. In a first approximation the inductance may be considered independent of the magnet's motion. The coil is supplied with a sinusoidal voltage with angular frequency ω. Find the expressions of active, reactive, and apparent powers as well as the *PF* of this system. Assume also some viscous damping that opposes the motion with a force proportional to the velocity of the mass.

2.8 A sinusoidal voltage $\mathbf{V_c} = 600\angle\delta$ V (phase angle δ can be adjusted in the range $0 < \delta < 360°$), is connected to an infinite bus $\mathbf{V} = 600\angle0°$ via a pure reactance $\omega L = 10 \ \Omega$. Sketch the phasor diagram and determine the geometric locus of the phasor $\mathbf{I} = (600 - 600\angle\delta)/_J 10$
Replace the reactance with a resistance $R = 10 \ \Omega$. Repeat the computations.

2.9 A load $\mathbf{Z_L} = 10\angle45°$ pu (per unit) is supplied by a voltage source $\mathbf{V_s} = 100\angle0°$ pu via a line with the impedance $\mathbf{Z_s} = 0.15 + j0.9$ pu. A static compensator with $X_{cM} = 0.94$ pu and adjustable voltage $\mathbf{V_c}\angle - \delta$ is used to control the reactive power flow. Your task is to observe the effect of the phase angle δ when adjusted in the range $0° \leq \delta \leq 20$ at four levels of normalized controlled voltage, $V_c/V_s = 1.00, \ 1.05, \ 1.07$ and 1.10. Graph the normalized load voltage V'/V versus δ using the base voltage $V = 92.91$ pu (this is the uncompensated load voltage.) These curves will prove the usefulness of such compensators not only as adjustable reactances but also as voltage regulators. Plot also the normalized compensator's active power P_c/P_L (the base power is the uncompensated load active power (2.44), $P_L = 616.63$ pu.) You will find that as the angle δ increases so does the converter's active power. Next focus on the compensator's reactive power, plot the normalized reactive power Q_c/Q_L ($Q_L = 616.63$ pu is the uncompensated load reactive power.) Lastly calculate and plot the overall power factor of the load and compensator.

2.10 Two parallel and identical lossless cylindrical conductors with a diameter $2a = 4 \times 10^{-3}$ m, are spaced $d = 0.1$ m apart. These conductors are located in the $x - z$ plane and supply with sinusoidal voltage $v = \sqrt{2}V \sin(\omega t)$ a load that is either a perfect resistance or a perfect inductance.
The magnetic field vector, perpendicular on the conductors' plane, at a point x m far from the center of the left conductor, is

$$\vec{H} = \left[\frac{i}{2\pi x} + \frac{i}{2\pi(d-x)}\right](-\vec{1}_y) = \frac{d}{2\pi(d-x)x} i \ (-\vec{1}_y)$$

The electric field vector in the conductor's plan is perpendicular on the conductors and has the expression[3]

$$\vec{E} = \frac{b}{2\ln[(b+d+2a)/(b-d+2a)]} \frac{1}{(d-x)x - a^2} v \ \vec{1}_x$$

---

[3] Kuffel E., Zaengl W. S.: "*High Voltage Engineering*," Pergamon Press, 1984, p.231.

where

$$b = \sqrt{d^2 - 4a^2}$$

Your task is to graph the Pointing vector distribution in the plane of the conductors. Plot the graphs $\wp(x) = |\vec{E} \times \vec{H}|$ for $\omega t = 0°$, $30°$, $45°$, $60°$, $90°$, $95°$, $120°$ and $135°$ and compare the energy flow patterns for the two types of load.

2.11 Study the transient that takes place when a circuit that consists of a resistance $R = 100\ \Omega$ in series with an inductance $L = 265.258$ mH is energized. The supply voltage is $v = 100\sin(2\pi 60t)$ V. Use any software package you are familiar with. Show that in steady-state the energy supplied to inductance is $w = W_{max}[1 + \cos(2\omega t)]$, where $W_{max} = LI^2$, $I = V/\sqrt{R^2 + \omega^2 L^2}$. Display the waveforms $v(t)$, $i(t)$, $v_L(t)$, $v(t)i(t)$ and the energies $\int_0^t v_L(t)i(t)\,dt$, $\int_0^t Ri(t)^2\,dt$, $\int_0^t v(t)i(t)\,dt$. Explain your results.

2.12 Add a capacitance $C = 26.526\ \mu$F in parallel with the inductance L in the above problem. This will cause parallel resonance. Repeat the study. Focus on the flow of energy into L, C and LC as a whole. Compare with the above problem and explain the differences.

2.13 A single-phase two-poles alternator (rarely seen today) supplies a resistance R. We assume this alternator as an ideal machine, linear, and lossless. The field winding carries the direct current I_F. The mutual inductance between the armature winding (stator) and field is $M = M_m \cos\theta$ where the angle $\theta = \omega t$, (note that for this two-poles machine $\omega = 2\pi f = 2\pi 60 = 376.9$ rad/s). The armature winding flux linkage is $\lambda = MI_F \cos(\omega t)$ and the induced voltage is

$$v = -d\lambda/dt = \omega M I_F \sin(\omega t)$$

If the leakage inductance is ignored we may approximate the supplied current as $i = v/R$. The prime mover provides a torque

$$T = T_e + J\frac{d\omega}{dt}$$

where J is the total momentum of inertia (alternator rotor plus prime mover), and $T_e = I_F i\, dM/d\theta$ is the electromagnetic torque.

The prime mover delivers a perfectly constant torque, $T = P/\omega_s$, to this lossless alternator. The alternator delivers the instantaneous power $p = P + p_i$, where $P = RI^2 = V^2/R$ and $p_i = -P\cos(2\omega t)$ is the intrinsic power. In steady-state this intrinsic power cannot be supplied by the prime mover, it is supported by the kinetic energy $J\omega^2/2$ stored in the rotating masses. This means that the rotor will turn with a velocity that is not constant and will have the expression

$$\omega = \omega_s + \Delta\omega(t); \quad \text{where} \quad \omega_s = 376.9 \text{ rad/s}$$

Your task now is to determine the expression of $\Delta\omega_s(t)$ for a 5 HP, 120 V, 60 Hz unit, that supplies rated power. The total momentum of inertia is $J = 0.5$ kgm^2.

2.14 A coaxial cable is designed to operate with a current density j A/m^2 and a maximum electric field $E = \sqrt{2}V/[a\ln(b/a)]$ V/m, where V is the rms voltage (assumed sinusoidal), a is the inner conductor external radius and b is the inner radius of the outer conductor. Find the mathematical expression that connects the radius b (hence, the size of the cable) with the apparent power $S = VI$, $I = \pi a^2 j$.

2.15 A single-phase load consists of a resistance R that varies from 25 Ω to 2500 Ω and it is supplied with a constant voltage $V = 105$ V, 60 Hz. The resistance is supplied from 105 V, 60 Hz via a feeder with the impedance $Z_S = 2 + j0.093$ Ω and a transformer with a magnetizing branch modeled by a shunt resistance $R_m = 3000$ Ω in parallel with a reactance $X_m = 12,000$ Ω. The equivalent short-circuit impedances are $\mathbf{Z_{Sp}} = \mathbf{Z_{Ss}} = 0.1 + j0.5$ Ω. (Z_{Sp}, Z_{Ss} are the primary and the secondary impedances). Plot the normalized power loss $\Delta P/\Delta P_0$ versus $(S/S_0)^2$ when R is varied from 25 to 2500 Ω. The base values ΔP_0 and S_0 are the total losses and the apparent power at $R = 50$ Ω. Compare the obtained graph with the one based on the approximation (2.36). In this case $a = 105^2/R_m$.

Note: The load voltage is kept constant $V = 105$ V rms, this means the load current $I = 105/R$. The source voltage is $\mathbf{V_S} = \mathbf{V} + \mathbf{Z_T I}$, where Z_T is the Thévenin impedance measured from the load's terminals. As R is reduced S increases and the voltage across the magnetizing branch increases causing the core losses to increase. This effect is causing a slight deviation from (2.36).

2.16 An uncompensated load with the voltage V, current I and power factor PF causes the line power losses $\Delta P = R_S I^2$. When the power factor is compensated to $PF' = 1.0$ the losses are reduced to $\Delta P'$. Prove that

$$PF = \sqrt{\frac{\Delta P'}{\Delta P}}$$

Note: This is an important expression, it defines the PF in a different manner. This expression should hold for any conditions, sinusoidal or nonsinusoidal, balanced or unbalanced, and demonstrates that the actual line utilization is not P/S, but $(P/S)^2$.

2.13   References

[1] Ghassemi F.: "A New Concept in ac Power Theory," IEE Proc. Generation, Transmission and Distribution, Vol. 147, No. 6, Dec. 2000, pp. 417–24.

[2] Stevens R. H.: "Power Flow Direction Definitions for Metering Bidirectional Power," IEEE Transactions on Power Apparatus and Systems, Vol. 102, No. 9, Sept. 1983, pp. 3018–21.

[3] Special AIEE Joint Committee: "Power Factor in Poly-Phase Circuits," AIEE, Vol. 39, July 1, 1920, pp. 1449–1520.

[4] Liénard A.: "À Propos de la Définition de la Puissance Reactive," (With Respect to the Definition of Reactive Power), Revue Generale d'Electricite, 22 May, 1926, p. 807.

[5] Curtis H. L., Silsbee F. B.: "Definitions of Power and related Quantities," AIEE Transactions, Vol. 54, No. 4, April 1935, pp. 394–404.

[6] American Institute of Electrical Engineers: "*American Standard Definitions of Electrical Terms.*" August 12, 1941, pp. 35–46.

[7] Richter R.: "Elektrische Maschinen," (Electrical Machines), Verlag Birkhauser, Basel, 1951.

[8] Emanuel A. E.: "Apparent Power Definitions for Three-Phase Systems," IEEE Transactions on Power Delivery, Vol. 14, No. 3, July 1999, pp. 767–72.

[9] Mohan N., Undeland T. M., and Robbins W. P.: "Power Electronics," John Wiley & Sons, 1995 (Chapter 18).

[10] Hingorani N. G., Gyugyi L.: "Understanding FACTS," IEEE Press, 2000.

[11] Acha E., Fuerte-Esquivel C. R., Ambriz-Perez H., and Angeles-Camacho C.: "FACTS Modeling and Simulation in Power Networks," John Wiley, 2004.

[12] Akagi H., Watanabe E. H., and Aredes M.: *"Instantaneous Power Theory and Applications to Power Conditioning,"* IEEE Press, Wiley–Interscience, 2007.

[13] Sen K. K., Sen M. L.: *"Introduction to FACTS Controller: Theory, Modeling and Applications,"* IEEE Press and John Wiley & Sons, 2009.

[14] Miller T. J. E.: *"Reactive Power Control in Electric Systems,"* John Wiley, 1982.

[15] Čakareski Ž., Emanuel A. E.: "Poynting Vector and the Quality of Electric Energy," European Trans. on Electrical Power, Vol. 11, No. 6, Nov/Dec 2001, pp. 375–81.

[16] Emanuel A. E.: "Poynting Vector and the Physical Meaning of Nonactive Powers," IEEE Transactions on Instrumentation and Measurement, Vol. 54, No. 4, August 2005, pp. 1457–62.

3

Single-Phase Systems with Nonsinusoidal Waveforms

Joy in looking and comprehending is nature's most beautiful gift.
—Albert Einstein, *Aphorisms for Leo Baeck*

This chapter addresses the periodic and nonsinusoidal conditions characterized by distorted voltage and current waveforms. In such cases the electric and magnetic fields surrounding the conductors are also nonsinusoidal. Since the waveforms are periodic the time variation of each wave can be expressed by means of a Fourier series. The harmonic voltages and currents are producing a multitude of Poynting vector components; some components are due to the interaction of a harmonic electric field E_m with a harmonic magnetic field H_n of the same order, $m = n = h$, and other are due to interactions among harmonics of different orders, $m \neq n$. The main goal of the following sections is to chart the flow of instantaneous powers and to understand their physical characteristics, peculiar to the nonsinusoidal regime.

A thorough knowledge of the nature of instantaneous powers, a clear view of their flow paths among sources and loads, or load to load, may come very handy when one tries to understand the impact a certain type of nonlinear load may have on a given power network, when the effectiveness of a filter has to be evaluated or when a set of power quality measurements, or recorded events, are to be analyzed.

This chapter starts with basic linear loads supplied with nonsinusoidal voltage and gradually progresses toward simple nonlinear loads – the actual sources of harmonics and the cause of voltage and current distortion – and finally analyzes a general case that provides the background needed for a critical evaluation of the different theories and approaches to the resolution of apparent power.

3.1 The Linear Resistance

A simple example will help us get acquainted with the components of instantaneous active power peculiar to nonsinusoidal conditions. We will assume an ideal resistor with resistance

Power Definitions and the Physical Mechanism of Power Flow   Alexander Eigeles Emanuel
© 2010 John Wiley & Sons, Ltd

R void of parasitic capacitance and inductance (this condition also implies no skin effect), supplied with a nonsinusoidal voltage[1].

$$v = v_1 + v_3 + v_5 = \widehat{V}_1 \sin(\omega t + \alpha_1) - \widehat{V}_3 \sin(3\omega t + \alpha_3) + \widehat{V}_5 \sin(5\omega t + \alpha_5) \qquad (3.1)$$

This voltage source can be viewed as an array of three voltage sources connected in series. This system being linear, the instantaneous current is found by using the superposition principle, thus

$$i = i_1 + i_3 + i_5 = \widehat{I}_1 \sin(\omega t + \alpha_1) - \widehat{I}_3 \sin(3\omega t + \alpha_3) + \widehat{I}_5 \sin(5\omega t + \alpha_5)$$

with $i_1 = v_1/R$, $i_3 = v_3/R$, $i_5 = v_5/R$, $\widehat{I}_1 = \widehat{V}_1/R$, $\widehat{I}_3 = \widehat{V}_3/R$ and $\widehat{I}_5 = \widehat{V}_5/R$.

The instantaneous power delivered to R is

$$\begin{aligned} p_p = vi &= (v_1 + v_3 + v_5)(i_1 + i_3 + i_5) = R(i_1 + i_3 + i_5)^2 \\ &= Ri_1^2 + R(i_3^2 + i_5^2) + 2R(i_1 i_3 + i_1 i_5 + i_3 i_5) \\ &= p_{p1} + p_{pH} \end{aligned}$$

where

$$p_{p1} = v_1 i_1 = Ri_1^2 = R\widehat{I}_1^2 \sin^2(\omega t + \alpha_1) = RI_1^2[1 - \cos(2\omega t + 2\alpha_1)] = P_1 + p_{i1}$$

We recognize that $P_1 = RI_1^2$ is the active power due to the interaction between the fundamental voltage and fundamental current, and $p_{i1} = -P_1 \cos(2\omega t)$ is the intrinsic power associated with the active power P_1.

The second term of p_p is due to harmonic currents i_3 and i_5 and has three subcomponents

$$p_{pH} = R(i_3^2 + i_5^2) + 2R(i_1 i_3 + i_1 i_5 + i_3 i_5) = P_H + p_{iH} + p_{iiH}$$

The first two components of p_{pH} are

$$P_H + p_{iH} = RI_3^2 + RI_5^2 - RI_3^2 \cos(6\omega t + 2\alpha_3) - RI_5^2 \cos(10\omega t + 2\alpha_5)$$

The constant terms are active powers

$$RI_3^2 + RI_5^2 = P_3 + P_5 = P_H$$

and the oscillating terms

$$-RI_3^2 \cos(6\omega t + 2\alpha_3) - RI_5^2 \cos(10\omega t + 2\alpha_5) = p_{i3} + p_{i5} = p_{iH}$$

are the intrinsic harmonic powers bound to the active powers P_3 and P_5.

---

[1] Everywhere in this book $\widehat{V}$ and $\widehat{I}$ represent voltage and current amplitudes or peak values, plain V and I are the respective rms values, and v and i are the instantaneous values.

The total power converted in heat by R is

$$P = P_1 + P_3 + P_5 = P_1 + P_H$$

The third and the last component in the expression of p_{pH} is

$$
\begin{aligned}
p_{iiH} &= 2R(i_1 i_3 + i_1 i_5 + i_3 i_5) \\
&= 4R[-I_1 I_3 \sin(\omega t + \alpha_1)\sin(3\omega t + \alpha_3) + I_1 I_5 \sin(\omega t + \alpha_1)\sin(5\omega t + \alpha_5) \\
&\quad - I_3 I_5 \sin(3\omega t + \alpha_3)\sin(5\omega t + \alpha_5)]
\end{aligned}
$$

This last term also has zero average value and is therefore a nonactive power. Moreover, it can be proved that just like the intrinsic powers p_{i1} and p_{iH}, the component p_{iiH} is not affecting the value of the rms current and is not causing power loss in the conductors that supply our load. If the supplying line has the resistance R_s, the power loss is

$$\Delta P = R_s I^2 = R_s \frac{V^2}{R^2} = \frac{R_s}{R} P = \frac{R_s}{R}(P_1 + P_H)$$

If we compute ΔP using equation (2.32) we find the same result,

$$\Delta P = \frac{R_s}{V^2} V^2 I^2 = \frac{R_s}{V^2} P^2 = \frac{R_s}{R}(P_1 + P_H)^2$$

confirming that p_{iiH} has no contribution to ΔP.

A fitting name for p_{iiH} may be *intrinsic power of the second-order.*

In the general case the nonsinusoidal voltage[2] impressed at the resistance's terminals is

$$v = \sum_h v_h = \sum_h \widehat{V}_h \sin(h\omega t + \alpha_h) = v_1 + v_H \qquad (3.2)$$

---

[2] In practical circuits not all of the harmonic frequencies are integer multiples of the power frequency, i.e. the harmonic of order h is not necessarily equal to 2, 3, 4, 5, ..., but in many situations the distorted waveforms contain frequency components called subharmonics ($0 \le h < 1$), and interharmonics (when $h > 1$, and h is not an integer number). The $h = 0$ component is the mean value and may be considered as a subharmonic with frequency $f_h = 0$. Modern documents recommend to avoid the name subharmonics and instead to use *sub-synchronous frequency components* or *sub-synchronous interharmonics*.

To indicate that all v components that are detected in the voltage and current spectra are included in a summation, many authors use the symbol $h \in v$; thus equation (3.2) may be written in a more rigorous mathematical form:

$$v = \sum_{h \in v} v_h = \sum_{h \in v} \widehat{V}_h \sin(h\omega t + \alpha_h) = v_1 + v_H$$

For the sake of clarity and avoidance of unnecessary heavy expressions the notation $h \in v$ will usually be avoided, nevertheless interharmonics and subharmonics are treated exactly like harmonics and, unless specified, such as $h = 1, 3, 5 \ldots, h$ includes all the measurable v components of the spectrum.

where

$$v_1 = \widehat{V_1} \sin(\omega t + \alpha_1) \tag{3.3}$$

is the fundamental instantaneous voltage and

$$v_H = \sum_{h \neq 1} \widehat{V_h} \sin(h\omega t + \alpha_h) \tag{3.4}$$

is the total instantaneous harmonic voltage. The load current is

$$i = \frac{v}{R} = \sum_h i_h = i_1 + i_H \tag{3.5}$$

with

$$i_1 = \widehat{I_1} \sin(\omega t + \alpha_1) \quad \text{and} \quad i_H = \sum_{h \neq 1} i_h = \sum_{h \neq 1} \widehat{I_h} \sin(h\omega t + \alpha_h) \tag{3.6}$$

(where $\widehat{I_1} = \widehat{V_1}/R$ and $\widehat{I_h} = \widehat{V_h}/R$), yielding the instantaneous power

$$p_p = vi = (v_1 + v_H)(i_1 + i_H) = v_1 i_1 + v_H i_H + v_1 i_H + v_H i_1 \tag{3.7}$$

The first term in (3.7)

$$p_{p1} = v_1 i_1 = \frac{v_1^2}{R} = P_1 + p_{i1} \tag{3.8}$$

is the *instantaneous fundamental power* (maybe it is more correct to name it *60 Hz or 50 Hz instantaneous power, when the power system's frequency is 60 or 50 Hz*) with the subcomponents

$$P_1 = V_1^2/R \tag{3.9}$$

the *fundamental active power* and

$$p_{i1} = -P_1 \cos(2\omega t + 2\alpha_1) \tag{3.10}$$

the *fundamental instantaneous intrinsic power*, described in section 2.1.

The second term in (3.7) is due to harmonics and has the following structure

$$v_H i_H = \sum_{h \neq 1} v_h i_h + \sum_{\substack{m \neq n \\ m,n \neq 1}} v_m i_n = p_{pH} + \sum_{\substack{m \neq n \\ m,n \neq 1}} v_m i_n \tag{3.11}$$

where

$$p_{pH} = P_H + p_{iH} + p_{iiH} \tag{3.12}$$

The constant term P_H, in (3.12), is *the total harmonic active power*

$$P_H = \sum_{h \neq 1} P_h; \quad P_h = V_h I_h = V_h^2/R = R I_h^2 \tag{3.13}$$

and the subcomponent P_h, in (3.13), is the *active power of the h-order harmonic*.

Always associated with P_H is the *instantaneous intrinsic harmonic power*

$$p_{iH} = -\sum_{h \neq 1} P_h \cos(2h\omega t + 2\alpha_h) = \sum_{h \neq 1} p_{ih} \tag{3.14}$$

where

$$p_{ih} = -P_h \cos(2h\omega t + 2\alpha_h) \tag{3.15}$$

is the *instantaneous intrinsic harmonic power of order h*.

The following complete array of instantaneous power terms are obtained after substituting (3.8) and (3.11) into (3.7):

$$p_p = vi = P_1 + p_{i1} + P_H + p_{iH} + p_{ii} \tag{3.16}$$

where

$$
\begin{aligned}
p_{ii} &= v_1 i_H + v_H i_1 + \sum_{\substack{m \neq n \\ m,n \neq 1}} v_m i_n = v_1 \sum_{h \neq 1} i_h + i_1 \sum_{h \neq 1} v_h + \sum_{\substack{m \neq n \\ m,n \neq 1}} v_m i_n \\
&= \frac{v_1}{R} \sum_{h \neq 1} v_h + \frac{v_1}{R} \sum_{h \neq 1} v_h + \frac{1}{R} \sum_{\substack{m \neq n \\ m,n \neq 1}} v_m v_n = \frac{1}{R} \sum_{\substack{m,n=1 \\ m \neq n}} v_m v_n \\
&= \frac{2}{R} \sum_{\substack{m,n=1 \\ m \neq n}} V_m V_n \sin(m\omega t + \alpha_m) \sin(n\omega t + \alpha_n)
\end{aligned}
\tag{3.17}
$$

is the *total instantaneous intrinsic power of second-order*.

Equation (3.2) indicates the possibility of using an equivalent circuit with two series connected voltage sources, Fig. 3.1. The superposition principle and the Kirchhoff's voltage law give:

$$v_1 = R i_1 \quad \text{and} \quad v_H = R i_H$$

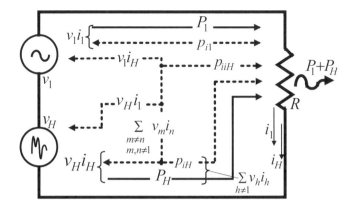

Figure 3.1 Linear resistance supplied with nonsinusoidal voltage: Power flow.

The fundamental voltage source v_1 supplies the instantaneous power

$$v_1(i_1 + i_H) = Ri_1^2 + Ri_1 i_H = P_1 + p_{i1} + v_1 i_H$$

and the harmonic voltage v_H supplies

$$v_H(i_1 + i_H) = v_H i_1 + P_H + p_{iH} + \sum_{\substack{m \neq n \\ m,n \neq 1}} v_m i_n$$

We see that v_H supplies the harmonic active power P_H, the harmonic intrinsic power p_{iH} and two of the three terms of the second-order intrinsic instantaneous power p_{ii} (3.17). The fundamental voltage v_1 supplies the fundamental active power, fundamental instantaneous intrinsic power plus the third term of p_{ii}.

The flow of all the instantaneous powers is shown in Fig. 3.1; solid unidirectional arrows indicate the flow of active powers P_1 and P_H. The bidirectional dashed lines show the intrinsic power flow. The intrinsic powers are inherently attached to the active powers, nevertheless, having no contribution to power loss in the line or to the power dissipated in the load, the intrinsic powers could be ignored or eliminated from Fig. 3.1 without losing consequential information.

3.2 The Linear Inductance

Let us look at an example similar to the one presented in the previous section. Now the load is a lossless inductance L supplied with the nonsinusoidal voltage v expressed in (3.1) (note the alternating signs). The steady-state current is

$$i = i_1 + i_3 + i_5 = -\widehat{I}_1 \cos(\omega t + \alpha_1) + \widehat{I}_3 \cos(3\omega t + \alpha_3) - \widehat{I}_5 \cos(\omega t + \alpha_5)$$

with $\widehat{I}_1 = \widehat{V}_1/X$, $\widehat{I}_3 = \widehat{V}_3/(3X)$ and $\widehat{I}_5 = \widehat{V}_5/(5X)$, $X = \omega L$.

This time the current waveform is not a replica of the voltage waveform and for this particular example its peak value is $\widehat{I} = \widehat{I}_1 + \widehat{I}_3 + \widehat{I}_5$. During each half-cycle the maximum value of the stored energy is

$$\frac{L\widehat{I}^2}{2} = \frac{L}{2}[\widehat{I}_1 + \widehat{I}_3 + \widehat{I}_5]^2 = L(I_1^2 + I_3^2 + I_5^2) + 2L(I_1 I_3 + I_1 I_5 + I_3 I_5)$$

This expression shows that the cross products $I_m I_n$ with $m \neq n$ also cause energy oscillations between L and v.

This phenomenon can be better understood if we express the voltage

$$v = L\frac{di}{dt} = L\left(\frac{di_1}{dt} + \frac{di_3}{dt} + \frac{di_5}{dt}\right)$$

yielding the instantaneous power

$$p_q = vi = Li\frac{di}{dt} = L\left(i_1\frac{di_1}{dt} + i_3\frac{di_3}{dt} + i_5\frac{di_5}{dt}\right)$$
$$+ L\left[i_1\left(\frac{di_3}{dt} + \frac{di_5}{dt}\right) + i_3\left(\frac{di_1}{dt} + \frac{di_5}{dt}\right) + i_5\left(\frac{di_1}{dt} + \frac{di_3}{dt}\right)\right] \quad (3.18)$$

When we compute the amount of energy stored in L, the reality of power oscillations of the type $Li_m di_n/dt$, that carry energy in and out of L, becomes more transparent:

$$w_L = \int_t^{t+\tau} p_q dt = L\int_t^{t+\tau} [i_1 di_1 + i_3 di_3 + i_5 di_5 + i_1(di_3 + di_5) + i_3(di_1 + di_5)$$
$$+ i_5(di_1 + di_3)]$$
$$= L\left[\sum_h \int_t^{t+\tau} i_h di_h + \sum_{\substack{m \neq n \\ m,n \neq 1}} \int_t^{t+\tau} i_m di_n\right]$$

The instantaneous power in our particular example is

$$p_q = vi = (v_1 + v_3 + v_5)(i_1 + i_3 + i_5) = v_1 i_1 + v_3 i_3 + v_5 i_5 + v_1 i_3 + v_1 i_5 + v_3 i_1$$
$$+ v_3 i_5 + v_5 i_1 + v_5 i_3$$
$$= X[-I_1^2 \sin(2\omega t + 2\alpha_1) - 3I_3^2 \sin(6\omega t + 2\alpha_3) - 5I_5^2 \sin(10\omega t + 2\alpha_5)$$
$$+ 2I_1 I_3 \sin(\omega t + \alpha_1)\cos(3\omega t + \alpha_3) - 2I_1 I_5 \sin(\omega t + \alpha_1)\cos(5\omega t + \alpha_5)$$
$$+ 6I_1 I_3 \sin(3\omega t + \alpha_3)\cos(\omega t + \alpha_1) + 6I_3 I_5 \sin(3\omega t + \alpha_3)\cos(5\omega t + \alpha_5)$$
$$- 10I_1 I_5 \sin(5\omega t + \alpha_5)\cos(\omega t + \alpha_1) + 10I_3 I_5 \sin(5\omega t + \alpha_5)\cos(3\omega t + \alpha_3)]$$

As expected this instantaneous power has zero average value and all the components are nonactive; moreover, comparing now the terms in the instantaneous power p_q expression, with the terms in the maximum energy expression, $L\widehat{I}^2/2$, we notice that the amplitude of every p_q term contributes to the instantaneous energy stored in L.

In the general case a nonsinusoidal current

$$i = \sum_h \widehat{I}_h \sin(h\omega t + \beta_h) \tag{3.19}$$

that flows through an inductance L will cause the voltage drop

$$v = L\frac{di}{dt} = \sum_h \widehat{V}_h \cos(h\omega t + \beta_h); \quad \widehat{V}_h = hX\widehat{I}_h; \quad X = \omega L \tag{3.20}$$

and the instantaneous power supplied to L has the expressions

$$p_q = vi = Li\frac{di}{dt} = \omega L \left\{ \sum_h h I_h^2 \sin(2h\omega t + 2\beta_h) \right.$$

$$\left. + \sum_{m \neq n} n I_m I_n \sin(m\omega t + \beta_m)\cos(n\omega t + \beta_n) \right\} \tag{3.21}$$

One may feel prompted to consider the second term in (3.21) as another intrinsic power component inherent to $Li\,di/dt$ in nonsinusoidal conditions. This idea may be reinforced by the fact that the rms current squared is $I^2 = \sum I_h^2$ and no $I_m I_n$ terms are found in the rms expression; however, when the active power loss in the supplying line resistance R_s is computed we find

$$\Delta P = R_s I^2 = \frac{R_s}{V^2} I^2 V^2 = \frac{R_s}{V^2} \sum_h I_h^2 \sum_h V_h^2 = \frac{R_s}{V^2}\left[\sum_h (hX I_h^2)^2 + \sum_{m \neq n} (nX I_m I_n)^2 \right] \tag{3.22}$$

proving that the currents cross products $I_m I_n$ contribute to power line losses.

The format used in (3.22) shows that the losses ΔP can be written as a function of all the amplitudes that define the power oscillations components present in p_q.

Using the same approach as in section 3.1 and starting with the voltage equations

$$v_1 = L\frac{di_1}{dt} \quad \text{and} \quad v_H = L\frac{di_H}{dt}$$

we find the instantaneous power supplied by the fundamental voltage source v_1 to be

$$v_1(i_1 + i_H) = Li_1\frac{di_1}{dt} + Li_H\frac{di_1}{dt}$$

and the instantaneous power supplied by the voltage source v_H to be

$$v_H(i_1 + i_H) = Li_1\frac{di_H}{dt} + Li_H\frac{di_H}{dt}$$

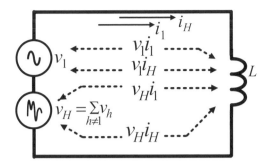

Figure 3.2 Linear inductance supplied with nonsinusoidal voltage: Instantaneous power flow.

These power-balance equations help to visualize the exchange of powers between the voltage source $v = v_1 + v_H$ and L as shown in Fig. 3.2.

3.3 The Linear Capacitance

The capacitance supplied by a nonsinusoidal voltage presents a situation similar to the conditions surrounding the inductance. The voltage

$$v = \sum_h \widehat{V}_h \cos(h\omega t + \beta_h) \tag{3.23}$$

applied to the terminals of an ideal linear capacitance C will supply the current

$$i = C\frac{dv}{dt} = -\sum_h \widehat{I}_h \sin(h\omega t + \beta_h); \quad \widehat{I}_h = Ch\omega\widehat{V}_h \tag{3.24}$$

The instantaneous power supplied by the source v is

$$p_q = vi = -\sum_h V_h I_h \sin(2h\omega t + 2\beta_h) + 2\sum_{m \neq n} V_m I_n \cos(m\omega t + \beta_m)\sin(n\omega t + \beta_n) \tag{3.25}$$

Evidently these power oscillations are nonactive. Energy is carried in and out of C by both terms in (3.25). The currents associated with these instantaneous powers cause line power loss as explained in the section 3.2, equation (3.22).

3.4 The Linear Series $R - L - C$ Circuit

We consider a nonsinusoidal voltage

$$v = \sum v_h = \sum_h \widehat{V}_h \sin(h\omega t + \alpha_h) \tag{3.26}$$

supplying a series $R - L - C$ circuit. In this case the current is

$$i = \sum_h i_h = \sum_h \widehat{I_h} \sin(h\omega t + \alpha_h - \theta_h); \quad \widehat{I_h} = \frac{\widehat{V_h}}{Z_h} \tag{3.27}$$

where

$$Z_h = \sqrt{R^2 + X_h^2}; \quad X_h = h\omega L - \frac{1}{h\omega C}; \quad \tan \theta_h = \frac{X_h}{R}$$

The details of the power flow mechanism become more evident if we separate the voltage v in two components

$$v = v_p + v_q = \sum_h v_{ph} + \sum_h v_{qh} \tag{3.28}$$

The first component includes every harmonic voltage v_{ph} in-phase with the harmonic current i_h. The second component, v_q, includes every harmonic voltage v_{qh} in-quadrature with the corresponding harmonic current i_h. For convenience we lump all the harmonic voltages in two groups: v_{pH} and v_{qH}, the total in-phase and in-quadrature harmonic voltages, i.e.

$$v_{pH} = \sum_{h \neq 1} v_{ph} \quad v_{qH} = \sum_{h \neq 1} v_{qh}$$

The fundamental instantaneous current i_1 and the total harmonic current $i_H = \sum_{h \neq 1} i_h$ have been introduced in (3.6) and can be used to express the load voltage (3.28) by means of four terms:

$$v = v_{p1} + v_{pH} + v_{q1} + v_{qH}$$

that lead to a linear equivalent circuit, originated from the superposition principle and sketched in Fig. 3.3. Each one of the four voltages provides a key equation:

$$v_{p1} = R i_1 \quad \text{and} \quad v_{q1} = L \frac{di_1}{dt} + \frac{1}{C} \int i_1 \, dt \tag{3.29}$$

for the fundamental voltages and current and

$$v_{pH} = R \sum_{h \neq 1} i_h \quad \text{and} \quad v_{qH} = L \sum_{h \neq 1} \frac{di_h}{dt} + \frac{1}{C} \sum_{h \neq 1} \int i_h \, dt \tag{3.30}$$

for the balance of harmonic voltages.

The instantaneous power

$$
\begin{aligned}
p = vi &= (v_{p1} + v_{pH} + v_{q1} + v_{qH})(i_1 + i_H) \\
&= (v_{p1} i_1) + (v_{pH} i_1 + v_{p1} i_H + v_{pH} i_H) + (v_{q1} i_1) + (v_{q1} i_H + v_{qH} i_1 + v_{qH} i_H) \tag{3.31}
\end{aligned}
$$

is separated in four components consequential to the grouping given in (3.29) and (3.30):

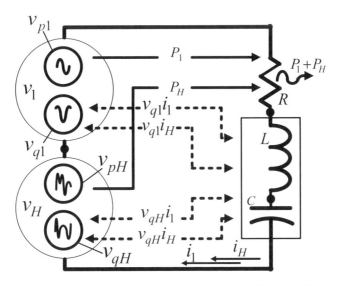

Figure 3.3 Linear R–L–C load supplied with nonsinusoidal voltage: The flow of instantaneous powers. (Intrinsic powers not shown).

The first component is

$$p_{p1} = v_{p1}i_1 = Ri_1^2 = RI_1^2[1 - \cos(2\omega t + 2\alpha_1 - 2\theta_1)] = P_1 + p_{i1} \qquad (3.32)$$

is the fundamental instantaneous power (active power and fundamental intrinsic power).

In the second group are included all the remaining components sustained by v_p:

$$v_{pH}i_1 = Ri_Hi_1 = R\sum_{h\neq1} i_h i_1$$

$$v_{p1}i_H = Ri_1i_H = R\sum_{h\neq1} i_1 i_h$$

$$v_{pH}i_H = R\sum_{h\neq1} i_h^2 + R\sum_{\substack{m\neq n \\ m,n\neq1}} i_m i_n$$

$$= R\sum_{h\neq1} I_h^2[1 - \cos(2h\omega t + 2\alpha_h - 2\theta_h)] + R\sum_{\substack{m\neq n \\ m,n\neq1}} i_m i_n = P_H + p_{iH} + R\sum_{\substack{m\neq n \\ m,n\neq1}} i_m i_n$$

The sum of these three components gives for the second term of p in (3.31):

$$p_{pH} = v_{pH}i_1 + v_{p1}i_H + v_{pH}i_H = P_H + p_{iH} + p_{iiH} \qquad (3.33)$$

where $p_{iiH} = R\sum_{m\neq n} i_m i_n$ is the second-order instantaneous intrinsic power.

The third term of p reveals the fundamental reactive power

$$p_{q1} = v_{q1}i_1 = Li_1\frac{di_1}{dt} + \frac{1}{C}i1\int i_1\,dt = V_{q1}I_1\sin(2\omega t + 2\alpha_1 - 2\theta_1) = Q_1\sin(2\omega t + 2\alpha_1 - 2\theta_1)$$

$$(3.34)$$

where $Q_1 = V_1 I_1 \sin\theta_1$.

The fourth term of p in (3.31) contains the remaining nonactive instantaneous power oscillating between the nonsinusoidal source and L–C. Three terms make this last component:

$$v_{q1}i_H = Li_H\frac{di_1}{dt} + \frac{i_H}{C}\int i_1\,dt = 2\sum_{h\neq1} V_{q1}I_h\cos(\omega t + \alpha_1 - \theta_1)\sin(h\omega t + \alpha_h - \theta_h)$$

$$v_{qH}i_1 = Li_1\frac{di_H}{dt} + \frac{i_1}{C}\int i_H\,dt = 2\sum_{h\neq1} V_{qh}I_1\cos(h\omega t + \alpha_h - \theta_h)\sin(\omega t + \alpha_1 - \theta_1) \quad (3.35)$$

and

$$v_{qH}i_H = Li_H\frac{di_H}{dt} + \frac{i_H}{C}\int i_H\,dt = \sum_{h\neq1} V_{qh}I_h\sin(2h\omega t + 2\alpha_1 - 2\theta_1)$$

$$+ 2\sum_{\substack{m\neq n \\ m,n\neq1}} V_{qm}I_n\cos(m\omega t + \alpha_m - \theta_m)\sin(n\omega t + \alpha_n - \theta_n) \quad (3.36)$$

This nonactive instantaneous power p_{qH} is the summation of the last three terms in (3.31):

$$p_{qH} = v_{q1}i_H + v_{qH}i_1 + v_{qH}i_H = \sum_{h\neq1} Q_h\sin(2h\omega t + 2\alpha_h - 2\theta_h)$$

$$+ 2\sum_{m\neq n} V_{qm}I_n\cos(m\omega t + \alpha_m - \theta_m)\sin(n\omega t + \alpha_n - \theta_n) \quad (3.37)$$

where $Q_h = V_{qh}I_h = V_hI_h\sin\theta_h$ is the amplitude of the oscillation with the frequency $2h\omega$, and $V_{qm}I_n = V_mI_n\sin\theta_m$ is the amplitude of a pair of oscillations with the frequencies $(m \pm n)\omega$. All the oscillations of the components included in p_{qH} are due to the charge-discharge of the L and C. Their nature is not different from the nature of the p_{q1} oscillation.

The instantaneous power flow paths are presented in Fig. 3.3.

In past literature [1] the terms Q_h were viewed as the components of a total reactive power $Q = \sum_h Q_h$, however, more recent documents [2] recognize Q_1 as a separate major component that gives significant information about the load performance.

3.5   The Nonlinear Resistance

Unlike the linear loads that cause the flow of a nonsinusoidal current only when supplied with nonsinusoidal voltage, the nonlinear loads will distort the current even when their input voltage is nearly sinusoidal. This means that nonlinear loads generate electric energy propagated by an electromagnetic field with frequencies different than the voltage source frequency.

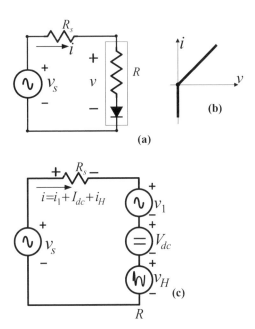

Figure 3.4 Nonlinear resistance: (a) Circuit. (b) v/i characteristic. (c) Equivalent circuit.

The study of the simplest of all the nonlinear loads, the nonlinear resistance, will help unravel the secret mechanism of harmonic power generation and gain a solid understanding of the types of instantaneous power sources peculiar to such circuits. The studied circuit shown in Fig. 3.4a has a nonlinear resistance that consists of a linear resistance R in series with an ideal diode D. The v/i characteristic is shown in Fig. 3.4b. The supply voltage is sinusoidal,

$$v_s = \sqrt{2}V_s \sin(\omega t) \tag{3.38}$$

nevertheless, the current i is distorted and has three components:

$$i = i_1 + I_{dc} + i_H \tag{3.39}$$

where

$$i_1 = \frac{V_s}{\sqrt{2}R_T} \sin(\omega t); \quad R_T = R_s + R \tag{3.40}$$

is the fundamental current,

$$I_{dc} = \frac{\sqrt{2}V_s}{\pi R_T} \tag{3.41}$$

is the direct current and

$$i_H = -\frac{2\sqrt{2}}{\pi}\frac{V_s}{R_T}\sum_{h=2,4,6..}\frac{\cos(h\omega t)}{h^2-1}\tag{3.42}$$

is the total harmonic current (all the harmonics, except the fundamental, are lumped together).

The distorted current causes a nonsinusoidal voltage drop across R_s, this causing in turn a nonsinusoidal voltage across the nonlinear load. Evidently in this case the spectrum of the nonlinear load voltage v has exactly the same harmonic orders as the current, therefore the nonlinear load's voltage has three components similar to (3.39):

$$v = v_1 + V_{dc} + v_h\tag{3.43}$$

This observation leads to an interesting idea: Let us replace the nonlinear load with a fictitious voltage v that has exactly the same spectrum as the actual voltage. In Fig. 3.4c we see the equivalent circuit that is used for this study. Based on Kirchhoff's law the three voltages are:

$$v_1 = v_s - R_s i_1 = \frac{V_s}{\sqrt{2}}\left(2 - \frac{R_s}{R_T}\right)\sin(\omega t)\tag{3.44}$$

$$V_{dc} = -R_s I_{dc} = \frac{-\sqrt{2}R_s}{\pi R_T}V_s\tag{3.45}$$

and

$$v_H = -R_s i_H = \frac{-2\sqrt{2}}{\pi}\frac{R_s}{R_T}V_s\sum_{h=2,4,6..}\frac{\cos(h\omega t)}{h^2-1}\tag{3.46}$$

The active power supplied by the source v_s is

$$P_s = \frac{1}{T}\int_o^T v_s i_1\,dt = V_s\frac{V_s}{2R_T} = \frac{V_s^2}{2R_T}\tag{3.47}$$

The fundamental current i_1 is causing the power loss

$$\Delta P_1 = R_s I_1^2 = R_s\left(\frac{V_s}{2R_T}\right)^2 = \frac{R_s}{4R_T^2}V_s^2\tag{3.48}$$

The fictitious fundamental voltage v_1 is absorbing the power

$$P_1 = \frac{1}{T}\int_o^T v_1 i_1\,dt = \frac{V_s}{2}\left(2 - \frac{R_s}{R_T}\right)\frac{V_s}{2R_T} = \frac{V_s^2}{2R_T}\left(1 - \frac{R_s}{R_T}\right)\tag{3.49}$$

The direct voltage V_{dc} generates the power

$$P_{dc} = V_{dc}I_{dc} = \frac{\sqrt{2}}{\pi}\frac{R_s}{R_T}V_s\frac{2\sqrt{2}V_s}{\pi R_T} = \frac{2R_s}{\pi^2 R_T^2}V_s^2 \tag{3.50}$$

The harmonic voltage source v_H generates energy that is dissipated in R_s. The active power supplied by v_H is

$$P_H = \frac{1}{T}\int_0^T v_H i_H \, dt = \frac{2\sqrt{2}}{\pi}\frac{R_s}{R_T}\frac{V_s}{\sqrt{2}}\frac{2\sqrt{2}}{\pi R_T}\frac{V_s}{\sqrt{2}}\sum_{h=2,4,\ldots}\frac{1}{(h^2-1)^2}$$

$$= \frac{4R_s}{\pi^2 R_T^2}V_s^2\sum_{h=2,4,\ldots}\frac{1}{(h^2-1)^2} = \frac{4R_s}{\pi^2 R_T^2}V_s^2\left(\frac{\pi^2}{16}-\frac{1}{2}\right) = \left(\frac{1}{4}-\frac{2}{\pi^2}\right)\frac{R_s}{R_T^2}V_s^2 \tag{3.51}$$

Now we should sort the above results on the base of the following three equations that apply to the circuit shown in Fig. 3.4c:

$$v_s = R_s i_1 + v_1 \tag{3.52}$$
$$V_{dc} = -R_s I_{dc} \tag{3.53}$$

and

$$v_H = -R_s i_H \tag{3.54}$$

Next we write the instantaneous powers supplied by each voltage

$$v_s i_1 = R_s i_1^2 + v_1 i_1 \tag{3.55}$$
$$V_{dc}I_{dc} = -R_s I_{dc}^2 \tag{3.56}$$

and

$$v_H i_H = -R_s i_H^2 \tag{3.57}$$

These three equations shed light on the following situation: The pure sinusoidal source v_s supplies power that covers a part of the losses dissipated by R_s and what remains enters v_1. The energy generated by V_{dc} and v_H is delivered entirely to R_s. This last observation leads the inquisitive reader to ask: "If the main source is sinusoidal and the generated energy is carried by an electromagnetic field with a frequency $f = 50$ or 60 Hz, how come there are present in this system electromagnetic waves with frequencies different than 50 or 60 Hz? In simpler terms, from where do these harmonic currents come? How are harmonics produced? The answer is found in the power conservation law. To facilitate the computation of the power terms involved in the power conservation equation, we need the squared rms current:

$$I^2 = \frac{1}{2\pi}\int_0^\pi i^2 \, d(\omega t) = \left(\frac{V_s}{\sqrt{2}R_T}\right)^2 \tag{3.58}$$

Next we can find the total power dissipated in R_s;

$$\Delta P = R_s I^2 = \frac{R_s}{2R_T^2} V_s^2 \tag{3.59}$$

and compare it with the sum of three powers: the power delivered by V_{dc} (3.50), by v_H (3.51), and the power supplied to R_s by v_s (3.48), i.e.

$$P_{dc} + P_H + \Delta P_1 = \frac{2R_s}{\pi^2 R_T^2} V_s^2 + \left(\frac{1}{4} - \frac{2}{\pi^2}\right) \frac{R_s}{R_T^2} V_s^2 + \frac{R_s}{4R_T^2} V_s^2 = \frac{R_s}{2R_T^2} V_s^2 = \Delta P \tag{3.60}$$

We found that all the energy generated by the nonfundamenal frequency sources is dissipated in R_s.

The nonlinear load converts energy into heat at the rate

$$P_{out} = RI^2 = (R_T - R_s)I^2 = \frac{R_T - R_s}{2R_T^2} V_s^2 = \left(1 - \frac{R_s}{R_T}\right) \frac{V_s^2}{2R_T} \tag{3.61}$$

This is exactly the power that will be measured by an ideal wattmeter connected at the nonlinear load's terminals.

Power enters the nonlinear load through v_1 and a part of it, P_{out}, is converted into heat and delivered to the surrounding media; the rest is delivered to V_{dc} and v_H.

Mathematically this explanation is easily proved by subtracting from the active power P_1, that enters the nonlinear load, the powers P_H and P_{dc}, delivered to the resistance R_s. Our explanation is correct, since the result of this subtraction equals the output power:

$$\begin{aligned} P_1 - P_H - P_{dc} &= \frac{V_s^2}{2R_T}\left(1 - \frac{R_s}{2R_T}\right) - \left(\frac{1}{4} - \frac{2}{\pi^2}\right)\frac{R_s}{R_T^2}V_s - \frac{2R_s}{\pi^2 R_T^2}V_s \\ &= \frac{V_s^2}{2R_T} - \frac{R_s}{4R_T^2}V_s^2 - \frac{R_s}{4R_T^2}V_s^2 = \left(1 - \frac{R_s}{R_T}\right)\frac{V_s^2}{2R_T} = P_{out} \end{aligned} \tag{3.62}$$

Equation (3.62) represents the conservation of active powers and is crucial information needed for the mapping of the active power flow sketched in Fig. 3.5. Probably the most exciting aspect of this display is the fact that according to (3.62) a portion of the power P_1 is converted in P_{dc} and P_H, i.e. $P_1 = P_{out} + P_{dc} + P_H$.

Besides instantaneous active and intrinsic powers there are also nonactive powers that oscillate between v_s and the nonlinear load in spite of the total absence of inertial components[3] from this circuit. These nonactive powers are recognized in the expression of the instantaneous power supplied by the source v_s:

$$p_s = v_s i = v_s i_1 + v_s I_{dc} + v_s i_H \tag{3.63}$$

---

[3] The inertial components are inductors, capacitors, and masses in motion belonging to electromechanical equipment.

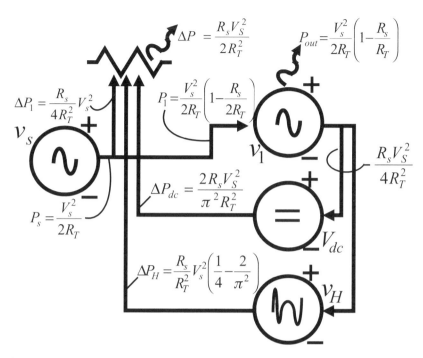

Figure 3.5   Active power flow in a noninductive circuit with a nonlinear resistance (replaced by three voltage sources v_1, V_{dc}, and v_H). (Intrinsic powers are not shown.)

The first term $v_s i_1$ is explained by (3.55): it is an instantaneous active power with its inherent instantaneous intrinsic powers (the phasors $\mathbf{V}_s$, $\mathbf{V}_1$ and $\mathbf{I}_1$ are in-phase.) Part of this active power is supplied to R_s and the remainder to the nonlinear resistance.

The remaining terms in (3.63) are nonactive, they are power oscillations with zero-mean values. The term $v_s I_{dc}$ can be deciphered starting from (3.52)

$$v_s I_{dc} = R_s i_1 I_{dc} + v_1 I_{dc} \tag{3.64}$$

From equation (3.53) we found

$$-R_s i_1 I_{dc} = V_{dc} i_1 \tag{3.65}$$

that substituted in (3.64) gives

$$v_s I_{dc} = -V_{dc} i_1 + v_1 I_{dc} \tag{3.66}$$

This result shows that the nonactive power oscillation $v_s I_{dc}$ is in balance with the nonactive power oscillations $V_{dc} i_1$ and $v_1 I_{dc}$. In the same way we deal with the instantaneous power $v_s i_H$. From (3.52) we find

$$v_s i_H = R_s i_H i_1 + v_1 i_H \tag{3.67}$$

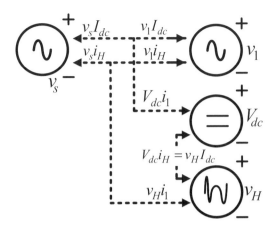

Figure 3.6 Nonactive power flow in a noninductive circuit with a nonlinear resistance. Power oscillations between the sinusoidal source and the nonlinear resistance (replaced by three voltage sources v_1, V_{dc}, and v_H).

and from (3.54) results

$$v_H i_1 = -R_s i_H i_1$$

that substituted in (3.67) gives

$$v_s i_H = -v_H i_1 + v_1 i_H$$

This group of oscillations takes place between v_s at one end of the connecting line and v_1 and v_H at the other end. These power oscillations also have zero mean values, none of them sustaining unidirectional flow of energy.

The last, but quite intriguing, category of oscillations takes place between the fictitious sources V_{dc} and v_H. From (3.53) and (3.54) results that

$$V_{dc} i_H = v_H I_{dc} = -R_s I_{dc} i_H$$

Thus, these oscillations are confined to the "space" found in the volume of the nonlinear load's mass, where the conversion from 60 or 50 Hz to dc and harmonic frequencies takes place. These particular oscillations are a link in the intricate chain that insures the conservation of instantaneous powers, Fig. 3.6.

3.6 The Nonlinear Inductance

In Fig. 3.7a is sketched the diagram of a simple circuit where a sinusoidal voltage $v_s = \widehat{V}_s \cos(\omega t)$ supplies a nonlinear inductance L_N connected in series with a linear inductance L. The nonlinear inductance has the flux linkage/current characteristic $\psi = \psi(i)$ shown in Fig. 3.7b; the permeability of the magnetic core material, $\mu \to \infty$ if $-\psi_s \le \psi \le \psi_s$ otherwise $\mu \approx 0$.

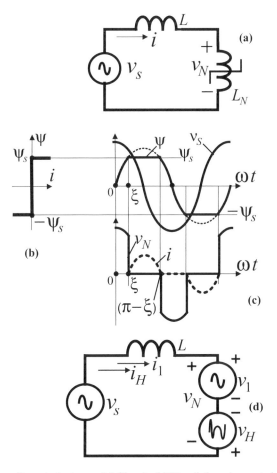

Figure 3.7 Lossless nonlinear inductance: (a) Circuit. (b) Flux linkage/current characteristic. (c) Waveforms. (d) Linear equivalent circuit.

The oscillograms of the voltage source v_s, flux linkage ψ, current i, and nonlinear inductance voltage v_N are presented in Fig. 3.7c.

During the interval $0 < \omega t < \xi$ the flux linkage $\psi < \psi_s$, the magnetizing current $i = 0$ and the nonlinear inductance voltage $v_N = v_s$. Also during this time the flux linkage has the expression

$$\psi = \int_0^t v_s \, dt = \frac{\widehat{V}_s}{\omega} \sin(\omega t)$$

At $\omega t = \xi$ the flux linkage reaches saturation, $\psi = \psi_s$, hence

$$\psi_s = \frac{\widehat{V}_s}{\omega} \sin(\xi) \quad \text{or} \quad \xi = \sin^{-1}\left(\frac{\omega \psi_s}{\widehat{V}_s}\right)$$

For the next interval $\xi \leq \omega t \leq (\pi - \xi)$ the flux remains constant, $\psi = \psi_s, v_N = d\psi/dt = 0$ and the current flow is governed by

$$v_s = L\frac{di}{dt} \quad \text{or} \quad i = \frac{1}{L}\int i \, dt = \frac{\widehat{V}_s}{\omega L}\sin(\omega t) + K$$

The integration constant is determined from the initial condition $i = 0$ at $\omega t = \xi$, thus $K = (\widehat{V}_s/\omega L)\sin(\xi)$ and the current has the expression

$$i = \frac{\widehat{V}_s}{\omega L}[\sin(\omega t) - \sin(\xi)]; \quad \xi \leq \omega t \leq (\pi - \xi)$$

Evidently this current is nonsinusoidal and has the following components

$$i = i_1 + i_H \quad i_1 = \widehat{I}_1\sin(\omega t) \quad i_H = \sum_{h\neq 1}\widehat{I}_h\sin(h\omega t)$$

The voltage across L_N has similar components, but leading the corresponding harmonic currents by $90°$;

$$v_N = v_1 - v_H \quad v_1 = \widehat{V}_1\cos(\omega t) \quad v_H = -\sum_{h\neq 1}\widehat{V}_h\cos(h\omega t)$$

The equivalent linear circuit is shown in Fig. 3.7d, (note the reverse polarities of v_1 and v_H). The Kirchhoff's voltage law applied to this circuit gives

$$v_s = L\frac{di_1}{dt} + v_1 \quad v_H = L\frac{di_H}{dt}$$

yielding the four instantaneous powers supplied by v_s and v_H:

$$v_s i_1 = Li_1\frac{di_1}{dt} + v_1 i_1 \quad v_s i_H = Li_H\frac{di_1}{dt} + v_1 i_H$$

$$v_H i_H = Li_H\frac{di_H}{dt} \quad\quad v_H i_1 = Li_1\frac{di_H}{dt}$$

Just like in the previous examples, an issue key to the understanding of i_H injection stems from the fact that the nonlinear inductance, assumed in this example, has no hysteresis and is lossless. Mathematically this translates in

$$v_N i = (v_1 - v_H)(i_1 + i_H) = 0 \quad \text{or} \quad v_1(i_1 + i_H) = v_H(i_1 + i_H)$$

The last equation[4] gives the following message: The instantaneous power absorbed by the fictitious fundamental voltage v_1 is converted in the instantaneous power generated by the

---

[4] One should not draw from this equation the conclusion that $v_1 = v_H$. The physical mechanism is more involving, the voltage v_1 is a fundamental, with a frequency of 50 or 60 Hz, and v_H contains all the higher harmonics.

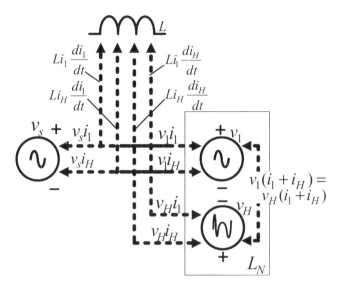

Figure 3.8 Instantaneous power flow in a circuit with a nonlinear lossless inductance. (See Fig. 3.7).

fictitious total harmonic voltage v_H. The instantaneous power flow pattern for the lossless nonlinear inductance is shown in Fig. 3.8.

A multitude of nonlinear circuits can be studied following the method described in sections 3.5 and 3.6. Among the problems listed at the end of this chapter the reader will find the lossy nonlinear inductance to be more challenging (see problem 3.8).

The most complete case is covered in the next section.

3.7   Nonlinear Load: The General Case

Many engineers list under the name of nonlinear loads all the categories of harmonic generating equipment, placing together actual nonlinear loads such as magnetic core devices (inductors, transformers and motors) and arcing devices (arc welders, arc furnaces, arc discharge lighting) with all sorts of loads such as converters (dc and ac adjustable speed drives, battery chargers, or any combination of the above.) The truth is that phase controlled loads, controlled or uncontrolled rectifiers, as well as cycloconverters, perform like a set of switches that periodically and sequentially connect and disconnect one or more linear, or nearly linear, loads to the supplying line. The correct name of such a load is *parametric*. For the purpose of this study, however, all harmonic generating loads are considered under one generic name of nonlinear loads.

In Fig. 3.9a is sketched a circuit where a nonlinear load NL is supplied from a network that has a Thévenin equivalent circuit that consists of a sinusoidal voltage

$$v_s = \widehat{V}_s \sin(\omega t)$$

in series with a resistance R_s and inductance L_s.

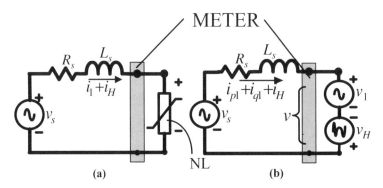

Figure 3.9 Nonlinear load; the general case: (a) Basic circuit. (b) Equivalent linear circuit.

The current is separated in fundamental and the total harmonic current:

$$i = i_1 + i_H \quad \text{where} \quad i_1 = \widehat{I_1} \sin(\omega t - \theta_1) \quad \text{and} \quad i_H = \sum_{h \neq 1} \widehat{I_h} \sin(h\omega t + \alpha_h) \quad (3.68)$$

The nonlinear load's voltage v can be represented by two equivalent voltages with opposite polarities, (Fig. 3.9b):

$$v = v_1 - v_H \quad \text{where} \quad v_1 = \widehat{V_1} \sin(\omega t) \quad \text{and} \quad v_H = \sum_{h \neq 1} \widehat{V_h} \sin(h\omega t + \alpha_h + \theta_h) \quad (3.69)$$

where the fundamental voltage phasor is taken as phase reference. Next, the current i_1 is divided into two components; an active current

$$i_{p1} = \widehat{I_1} \cos \theta_1 \sin(\omega t) \quad (3.70)$$

in-phase with the voltage v_1, and a reactive component

$$i_{q1} = -\widehat{I_1} \sin \theta_1 \cos(\omega t) \quad (3.71)$$

in-quadrature with v_1.

The reason for this separation of currents stems from the need to single out the fundamental instantaneous active power measured at the NL terminals.

The Kirchhoff's voltage law for the fundamental frequency is

$$v_s = R_s i_1 + L_s \frac{di_1}{dt} + v_1 \quad (3.72)$$

therefore the fundamental instantaneous power supplied by v_s is

$$v_s i_1 = v_s(i_{p1} + i_{q1}) = R_s i_1(i_{p1} + i_{q1}) + L_s(i_{p1} + i_{q1})\frac{di_1}{dt} + v_1 i_{p1} + v_1 i_{q1} \quad (3.73)$$

The nonactive component resulting from the interaction of v_s with i_H is

$$v_s i_H = R_s i_1 i_H + L_s i_H \frac{di_1}{dt} + v_1 i_H \tag{3.74}$$

The fundamental instantaneous active power is

$$p_{p1} = v_1 i_{p1} = \widehat{V}_1 \sin(\omega t)[\widehat{I}_1 \cos\theta_1 \sin(\omega t)] \tag{3.75}$$

Similarly one finds the fundamental instantaneous reactive power

$$p_{q1} = v_1 i_{q1} = \widehat{V}_1 \sin(\omega t)[-\widehat{I}_1 \sin\theta_1 \cos(\omega t)] \tag{3.76}$$

From (3.72) we find that

$$v_s i_{p1} = R_s i_1 i_{p1} + L_s i_{p1} \frac{di_1}{dt} + p_{p1} \tag{3.77}$$

and

$$v_s i_{q1} = R_s i_1 i_{q1} + L_s i_{q1} \frac{di_1}{dt} + p_{q1} \tag{3.78}$$

The Kirchhoff's voltage law written for the harmonic voltages is

$$v_H = \sum_{h \neq 1} v_h = R_s \sum_{h \neq 1} i_h + L_s \sum_{h \neq 1} \frac{di_h}{dt} = R_s i_H + L_s \frac{di_H}{dt} \tag{3.79}$$

this giving the instantaneous power due to i_1

$$v_H i_1 = R_s i_H i_1 + L_s i_1 \frac{di_H}{dt} \tag{3.80}$$

and the instantaneous power due to i_H

$$v_H i_H = \sum_{h \neq 1} v_h \sum_{h \neq 1} i_h = \sum_{h \neq 1} v_h i_h + \sum_{\substack{m \neq n \\ m,n \neq 1}} v_m i_n$$

$$= R_s \sum_{h \neq 1} i_h^2 + R_s \sum_{\substack{m \neq n \\ m,n \neq 1}} i_m i_n + L_s \left[\sum_{h \neq 1} i_h \frac{di_h}{dt} + \sum_{\substack{m \neq n \\ m,n \neq 1}} i_m \frac{di_n}{dt} \right] \tag{3.81}$$

The first term in (3.81) is a part of the power supplied to R_s, including the total harmonic active power P_H and its instantaneous intrinsic power of the first-order. We recognize the second term in (3.81) as a part of the *instantaneous intrinsic power of mn-order*. The last term represents power oscillations between v_H and L_s.

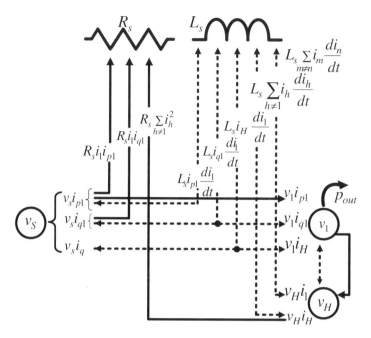

Figure 3.10 Flow of instantaneous powers that replicate the flow of Poynting vector components. (Intrinsic power flow is not included).

Next we check on the instantaneous power $v_s i_H$. From (3.72) results

$$v_s i_H = R_s i_1 i_H + L_s i_H \frac{di_1}{dt} + v_1 i_H \tag{3.82}$$

Both first terms in (3.80) and (3.82) are part of the intrinsic instantaneous powers p_{iiH} and are left out of the instantaneous power flow map presented in Fig. 3.10.

The set of equations (3.74) to (3.82) helped to determine the chart of instantaneous power flow as depicted in Fig. 3.10. As in the previous examples we found that part of the active power supplied by v_s goes to R_s and the rest to v_1. Some active power is also supplied by v_H to R_s.

The voltage drop across the linear inductance L_s is

$$v_{Ls} = L_s \frac{di}{dt} = L_s \left(\frac{di_1}{dt} + \frac{di_H}{dt} \right)$$

and the instantaneous power supplied to L_s is

$$p_{Ls} = v_{Ls}(i_1 + \sum_{h \neq 1} i_h) = L_s \left(i_1 \frac{di_1}{dt} + i_1 \frac{di_H}{dt} + i_H \frac{di_1}{dt} + \sum_{h \neq 1} i_h \frac{di_h}{dt} + \sum_{\substack{m \neq n \\ m,n \neq 1}} i_m \frac{di_n}{dt} \right)$$

The sinusoidal voltage source v_s provides three instantaneous powers: $v_s i_{p1}$ (3.77), $v_s i_{q1}$ (3.78) and $v_s i_H$ (3.82). The nonlinear load NL receives instantaneous active power via $v_1 i_{p1}$ (3.75) and delivers instantaneous active power via the active components of $v_H i_H$ (3.81).

This nonlinear load is interacting with the energy supplied by five distinct instantaneous powers: The power $v_1 i_{p1}$ a unidirectional instantaneous power. The component $v_1 i_{q1}$ an oscillation with a nil mean value involving L_s, NL and v_s. In the same category are the powers $v_1 i_H$ and $v_H i_1$. Lastly the instantaneous powers $v_H i_H$ that contain unidirectional components flowing from NL toward R_s and instantaneous powers oscillating between NL and L_s.

Assuming that the nonlinear load delivers the instantaneous power p_{out}, the energy conservation law requires that

$$(v_1 - v_H)(i_1 + i_H) = p_{out}$$
$$\text{or } v_1 i_1 + v_1 i_H = p_{out} + v_H i_1 + v_H i_H \tag{3.83}$$

The puzzle is completed using (3.83) that points to the mechanism of harmonics generation:

$$\int_0^T (v_1 i_1 + v_1 i_H) \, dt = \int_0^T p_{out} \, dt + \int_0^T (v_H i_1 + v_H i_H) \, dt$$

This equation reads: "The electric energy supplied to a nonlinear load equals the energy converted by the load in heat, or in other forms of energy, plus energy that is converted in energy at higher harmonics frequencies and returned to the network."

The beauty of the instantaneous power flow map, shown in Fig. 3.10, is its very agreement with the flow of energy as described by means of Poynting vector. Each instantaneous power shown in Fig. 3.10 is mirrored in one of the components of the electromagnetic wave with the energy density flow rate $\vec{\wp} = \vec{E} \times \vec{H}$. This claim is based on the fact that the Poynting vector is proportional to the instantaneous power, $\wp = (K_E K_H / r^2) v i$, (2.55).

According to Fig. 3.10 a specially designed power quality instrument or power meter, connected at the nonlinear load terminals, will detect the following above mentioned five instantaneous powers:

From (3.75) results

$$p_{p1} = v_1 i_{p1} = V_1 I_1 \cos \theta_1 [1 - \cos(2\omega t)] = P_1 + p_{i1}; \quad P_1 = V_1 I_1 \cos \theta_1 \tag{3.84}$$

and from (3.76)

$$p_{q1} = -Q_1 \sin(2\omega t); \quad Q_1 = V_1 I_1 \sin \theta_1 \tag{3.85}$$

Sustained by $v_s i_H$ in (3.74) is "hidden" the *instantaneous current distortion power*:

$$p_{DI} = v_1 i_H = 2V_1 \sum_{h \neq 1} I_h \sin(\omega t) \sin(h\omega t + \alpha_h)$$
$$= V_1 \sum_{h \neq 1} I_h \{\cos[(h - 1)\omega t + \alpha_h] - \cos[(h + 1)\omega t + \alpha_h - \theta_h]\} \tag{3.86}$$

From equation (3.80) one finds the *instantaneous voltage distortion power* produced by $v_H i_1$, that covers the oscillations $L_s i_1 \, di_H/dt$. The detailed expression of these oscillations is

$$p_{DV} = v_H i_1 = 2I_1 \sum_{h \neq 1} V_h \sin(\omega t - \theta_1) \sin(h\omega t + \alpha_h + \theta_h)$$

$$= I_1 \sum_{h \neq 1} V_h \{\cos[(h-1)\omega t + \alpha_h + \theta_1 + \theta_h] - \cos[(h+1)\omega t + \alpha_h - \theta_h - \theta_1]\} \quad (3.87)$$

Lastly, from (3.81) we obtain the *instantaneous harmonic power*:

$$p_H = v_H i_H = R_s \sum_{h \neq 1} [\widehat{I}_h \sin(h\omega t + \alpha_h + \theta_h)]^2 + R_s \sum_{\substack{m \neq n \\ m,n \neq 1}} i_m i_n$$

$$+ L_s \omega \sum_{h \neq 1} h \widehat{I}_h^2 \sin(h\omega t + \alpha_h) \cos(h\omega t + \alpha_h)$$

$$+ L_s \omega \sum_{\substack{m \neq n \\ m,n \neq 1}} n \widehat{I}_m \widehat{I}_n \sin(m\omega t + \alpha_m) \cos(n\omega t + \alpha_n)$$

$$= \sum_{h \neq 1} (P_h + p_{ih}) + R_s \sum_{\substack{m \neq n \\ m,n \neq 1}} i_m i_n + \sum_{h \neq 1} Q_h \sin(2h\omega t + 2\alpha_h)$$

$$+ 2L_s \omega \sum_{\substack{m \neq n \\ m,n \neq 1}} n I_n I_m \{\sin[(m-n)\omega t + \alpha_m - \alpha_n]$$

$$- \sin[(m+n)\omega t + \alpha_m + \alpha_n]\} \quad (3.88)$$

supplied by the harmonic voltage v_H. These terms cover the total harmonic power $P_H = \sum_{h \neq 1} P_h$, the instantaneous intrinsic power $p_{iH} = \sum_{h \neq 1} p_{ih}$, a part of the intrinsic power of second-order, and the power oscillations between L_s and v_H.

Only the instantaneous powers p_{p1} and p_H have average values and provide the active powers P_1 and P_H, respectively. All the remaining instantaneous powers are nonactive.

From Fig. 3.10 results that the instantaneous power line loss is

$$\Delta p = R_s \left(i_1 i_{p1} + i_1 i_{q1} + \sum_{h \neq 1} i_h^2 \right) \quad (3.89)$$

Substitution of $i_1 = i_{p1} + i_{q1}$ and $i_h = i_{ph} + i_{qh}$ in (3.89) yields

$$\Delta p = R_s \left[i_{p1}^2 + i_{q1}^2 + 2i_{p1} i_{q1} + \sum_{h \neq 1} (i_{ph}^2 + i_{qh}^2 + 2i_{ph} i_{qh}) \right]$$

therefore the active power lost in the line that supplies the nonlinear load is

$$\Delta P = R_s \left[I_{p1}^2 + I_{q1}^2 + \sum_{h \neq 1}(I_{ph}^2 + I_{qh}^2) \right]$$

$$= R_s \left\{ \left(\frac{P_1}{V_1}\right)^2 + \left(\frac{Q_1}{V_1}\right)^2 + \sum_{h \neq 1}\left[\left(\frac{P_h}{V_h}\right)^2 + \left(\frac{Q_h}{V_h}\right)^2\right] \right\} \qquad (3.90)$$

This result is important since it proves that the power loss in the supplying line that carries nonsinusoidal currents is a function of the active powers squared (P_1^2, $\sum_{h \neq 1} P_h^2$), reactive powers squared, (Q_1^2, $\sum Q_h^2$) and the voltages V_1^2 and V_h^2.

The expression (2.32), advocated in Chapter 2 for the sinusoidal situations, can be applied as well to single-phase nonsinusoidal conditions:

$$\Delta P = R_s I^2 = \frac{R_s}{V^2}V^2 I^2 = \frac{R_s}{V^2}\left[V_1^2 + \sum_{h \neq 1}V_h^2\right]\left[I_{p1}^2 + I_{q1}^2 + \sum_{h \neq 1}(I_{ph}^2 + I_{qh}^2)\right]$$

$$= \frac{R_s}{V^2}\left[P_1^2 + Q_1^2 + V_1^2\sum_{h \neq 1}I_h^2 + I_1^2\sum_{h \neq 1}V_h^2 + \sum_{h \neq 1}V_h^2\sum_{h \neq 1}I_h^2\right] \qquad (3.91)$$

This last expression is significant since it shows that all five forms of instantaneous powers p_{p1}, p_{q1}, p_{DI}, p_{DV}, and p_H, ((3.84) to (3.87)), affect the power lost in the supplying line.

If we consider the ac motors and review the theory of rotating field, we realize that each harmonic voltage produces a rotating magnetic flux. However, it is only the fundamental voltage that supports a useful rotating magnetic field; the remaining harmonic rotating fields are detrimental, causing parasitic torques and additional losses. Keeping in mind that ac motors make the majority of loads (true for three-phase systems), it is justified to view the harmonic power, active and nonactive, as a by-product of the conversion of 60/50 Hz energy into harmonic frequency energy and to consider harmonic energy generated by the nonlinear loads as *electromagnetic pollution*.

The harmonic active power $P_H = \sum_{h \neq 1} P_h = \sum_{h \neq 1} V_h I_h \cos\theta_h$ is considered by many engineers as a reasonable indicator of harmonic pollution. In the vast majority of practical situations $P_H \ll P_1$. The minute value of each P_h is due to the fact that the phase angle $|\theta_h| \approx 90°$ ($\tan\theta_h = h\omega L_s/R_s$ and usually $h\omega L_s \gg R_s$). If the load is linear $P_H > 0$, if the load is nonlinear and generates harmonic energy, then $P_H < 0$. This is true only if the nonlinear load is large enough or decoupled from other nonlinear loads. There are many situations when a nonlinear load absorbs energy at certain harmonic frequencies and delivers energy at other frequencies [3,4]. This happens when the observed nonlinear load is dominated by other nonlinear loads supplied by the same feeder and the observed nonlinear load acts as a sink (an active filter). This means that one or more harmonic current phasors injected by the observed load are out of phase with the larger phasors injected by the larger loads. (See problems 3.14 and 3.15.)

The analyses, results, and observations presented in this chapter may seem detached from the actual engineering design of a metering instrument, the selection of switchgear, or planning an optimal mitigation strategy for power factor improvement. Nevertheless, this material provides a solid background conducive to an objective evaluation of different apparent power resolutions and helps to establish the merits and the shortcomings of different schools of thought.

3.8 Problems

3.1 A single-phase line consists of two resistances connected in series. From left to right $R_1 = 10 \, \Omega$ and $R_2 = 90 \, \Omega$. The left-end of this line is supplied with a voltage $v_{s1} = 400 \sin(\omega t)$ and the right-end with a voltage $v_{s5} = 800 \sin(5\omega t + 60°)$. The resistances are connected at the node A. The return line has zero-resistance and is considered node B. Compute and sketch the flow of instantaneous powers. Compute the instantaneous and the active power supplied through the terminals A B.

3.2 Repeat problem 3.1 when $R_1 = 20 \, \Omega$ and $R_2 = 80 \, \Omega$. Explain the differences between the two results.

3.3 A linear resistance $R = 1 \, \Omega$ is supplied with a voltage

$$v = \sqrt{2}[5 \sin \omega t + 4 \sin(3\omega t) + 3 \sin(5\omega t + \pi/6)]$$

Compute the fundamental active power P_1, the total harmonic active power P_H, and the apparent power $S = VI = P_1 + \sum_{h \neq 1} P_h$. Check the equality

$$\left[P_1 + \sum_{h \neq 1} P_h \right]^2 = P_1^2 + \sum_{h \neq 1} P_h^2 + P_D^2$$

where $P_D^2 = \sum_{m \neq n}(P_m P_n)^2; m, n = 1, 3, 5$.

3.4 An inductance $L = 10$ H is supplied with $v = \sqrt{2}[\sin(\omega t) + \sin(3\omega t + \theta_3)]$, $\omega = 2\pi$ rad/s, $f = 1$ Hz. Using your preferred software package compute and plot the maximum energy stored in L, $W_{max} = W_{max}(\theta_3)$ for $0 \leq \theta_3 \leq 360°$. Display the waveforms $v = v(t)$, $i = i(t)$, $p(t) = v(t)i(t)$ and $\int p(t)dt$ for $\theta_3 = 135°$.

3.5 Prove that for an ideal linear resistor (not affected by skin effect and free from parasitic components) supplied with a nonsinusoidal voltage with a total harmonic distortion THD_V, the ratio $P_1/S = 1/[1 + THD_V^2]$.
Note: The total harmonic distortion $THD_V = THD_I = \sqrt{\sum_h (V_{h \neq 1}/V_1)^2}$ and $S = VI$.

3.6 Compute the harmonic currents and voltages for the lossless nonlinear inductance described in section 3.6. Assume

$$v = \sqrt{2} \, 120 \sin(376.99t) \text{ V and } \quad \xi = 30°$$

The line reactance is $\omega L = 376.99\,L = 10\,\Omega$. Complete the instantaneous power flow map. (You may use Matlab, PSpice, or any adequate software to solve this problem.)

3.7 Repeat problem 3.6 with the line inductance replaced by a 10 Ω resistance.

3.8 Repeat problem 3.6 if the lossless inductance is replaced with a lossy inductance. Assume the ψ/i characteristic to have a rectangular hysteresis loop, such that when $\psi < |\psi_s|$ $i = |I_0|$. If $d\psi/dt > 0$ then $i = I_0$. If $d\psi/dt < 0$, $i = -I_0$. When $\psi = \psi_s$, $-I_0 < i < \infty$. When $\psi = -\psi_s$ $-\infty < i < I_0$. The hysteresis power loss is $\Delta P_{hyst} = 4\psi_s I_0 f$, $f = 60$ Hz. Assume $I_0 = 2$ A.

3.9 A direct voltage source, $V_s = 100$ V, supplies a mechanical switch via a 10 Ω resistance. The mechanical switch, considered a time-varying load, is turned on/off with a 50% duty cycle, (i.e., 1 ms is closed and 1 ms is kept open.) You can easily determine the Fourier series for the current through and the voltage across the switch and figure the linear equivalent circuit. In reference [5] it is claimed that no energy flows and no energy oscillations take place between the switch and the source. Using the method studied in this chapter, determine the flow of all the instantaneous powers characteristic to this circuit. Make a rough sketch of the Poynting vector flux lines patterns for $f = 0$ and $f = 1.0$ kHz.

3.10 Repeat problem 3.9 if 8 Ω of the 10 Ω resistance are included in the load, (this leaves 2 Ω in the supplying line.)

3.11 A sinusoidal voltage $v_s = \widehat{V}_s \sin(\omega t)$ supplies a rectifier bridge via a resistance R_s. The four diodes are considered ideal (zero voltage drop during conduction, infinite resistance when reverse biased.) The dc load is represented by a current source I_0. Due to resistance R_s there is a commutation time when all four diodes conduct, (as long as $|i| \le I_0$, the bridge acts like a short-circuit and the line current $i = v_s/R_s$).

During the intervals $k(\pi \pm \xi) < \omega t < (k+1)(\pi \pm \xi)$, $k = 1, 2, 3 \ldots$, the line current $|i| = I_0$. When $\omega t = \xi$ $i = I_0 = (\widehat{V}_s/R_s) \sin(\xi)$.

Your task is to determine the components of the instantaneous power that supplies the rectifier bridge. Find the expressions of v_1, v_H, i_1 and i_H and work your way to a complete map of instantaneous powers. Prove that $P_0 = V_{dc} I_0 = P_1 - P_H$.

3.12 Sketch the instantaneous power, flow map for a linear R–L–C load supplied by a nonsinusoidal voltage. Base your model on fundamental and harmonic current components in-phase and in-quadrature with the fundamental and harmonic voltages supplying the load.

3.13 Obtain the power flow map for the general case of a nonlinear load. Take the fundamental active current $i_1 = \widehat{I}_1 \sin(\omega t)$ as reference. Separate the nonlinear load voltage in an active component $v_{p1} = \widehat{V}_1 \cos\theta_1 \sin(\omega t)$ and a reactive component $v_{p1} = \widehat{V}_1 \sin\theta_1 \sin(\omega t)$. It may be necessary to separate also the supply voltage v_s in active and reactive components v_{sp} and v_{sq}. Compare your map with Fig. 3.10. You will observe that the new approach is useful when the recording instrument is connected at the terminals of v_s.

3.14 A current harmonic source $i_h = \sin(h\omega t)$ is connected at the terminals A and C into a circuit that has $R_{AB} = 8\,\Omega$ connected node A to node B and $R_{BC} = 2\,\Omega$ connected node B to node C. The current flows from C to A.

Compute the harmonic voltage $v_h = v_{AC}$ and the instantaneous and the active power supplied by the current source i_h. Next a larger current source, $i'_h = 10 \sin(h\Omega t)$ $(\Omega \ne \omega)$, is connected at the terminals BC (the current flows from B to C). Repeat the calculations and comment on the obtained results.

3.15 Two controlled rectifiers, A and B, are connected in parallel at the same bus. Both rectifiers supply zero-ripple dc current to their respective dc loads. We label these dc currents I_A and I_B. If the commutation phenomenon is ignored then the ac line currents are perfect square waves with the expressions

$$i_A = \frac{4I_A}{\pi} \sum_{h=1,3,5,...} \frac{\sin(h\omega t - h\alpha_A)}{h} \quad \text{and} \quad i_B = \frac{4I_B}{\pi} \sum_{h=1,3,5,...} \frac{\sin(h\omega t - h\alpha_B)}{h}$$

Compute and plot P_{Bh}/P_{Ah} versus α_B for $0 < \alpha_B < 90°$. Take the ratio $0.1 < I_B/I_A < 0.9$ as parameter. Assume $\alpha_A = 0$. Complete your study for $h = 3$ and $h = 5$. Ignore supplying line inductance. Consider line resistance R_s. Assume a sinusoidal voltage source, $\widehat{V} \sin(\omega t)$ and $R_s I_A = 0.05\widehat{V}$.

3.9   References

[1] IEEE Std 100, *"IEEE 100, The Authoritative Dictionary of IEEE Standards Terms,"* 7th Edition, 2000, IEEE Press.

[2] IEEE Std 1459–2010, *"IEEE Standard Definitions for the Measurement of Electric Power Quantities Under Sinusoidal, Nonsinusoidal, Balanced or Unbalanced Conditions"*. (Upgraded to Full-Use, August 2002.)

[3] Emanuel A. E.: "Harmonic Generation Modeling and Harmonic Power Flow in Power Systems," *Proceedings of the Second International Conference on Harmonics in Power Systems*, Manitoba, Oct. 1986, pp. 158–63.

[4] Emanuel A. E.: "On the Assessment of Harmonic Pollution," *IEEE Trans. on Power Delivery*, vol. 10, No.3, July 1995, pp. 1693–98.

[5] Czarnecki L. S.: *"Harmonics and Power Phenomena,"* Wiley Encyclopedia of Electrical and Electronics Engineering. Supplement 1, pp. 195–218. Editor: John G. Webster. Wiley 2000.

4

Apparent Power Resolution for Nonsinusoidal Single-Phase Systems

There is nothing so powerful as truth; and often nothing so strange.
—Daniel Webster, *April 1830*

Let us consider a modern spectrum, or wave-analyser, connected at the terminals of a load, linear or nonlinear, that accurately measures the v harmonic voltage phasors and the v harmonic current phasors that define the respective voltage and current waves. The interaction among these harmonic voltages and currents of the same frequency yields v instantaneous active powers and v instantaneous nonactive powers. Moreover, the cross products of voltage and current harmonics of different frequencies yield $v(v-1)$ more instantaneous nonactive powers. With so many elementary powers, (a total of $v^2 + v$) it is easy to imagine a multitude of possible resolutions of the apparent power S in seemingly useful components separation. No wonder that the engineering literature in the last eight decades abounds with papers, each one describing an improved, or a novel resolution of S, claiming to be more practical than the last due to regrouping of the powers according to a clearer physical meaning or a more refined economical analysis.

The resolution of apparent power for nonsinusoidal conditions was from the beginning a controversial topic that caused, and is still causing today, passionate discussions. In 1933 A. E. Knowlton chaired the famous AIEE Schenectady meeting that turned into one of the most heated debates in the history of electrical engineering [1]. The discussions were fueled by a set of papers presented by the AIEE elite: C. L. Fortescue, V. G. Smith, W. V. Lyon and W. H. Pratt. There are 21 recorded discussions. A noteworthy comment was then made by V. Karapetoff:

"Any definition of power factor that can not be realized with fairly simple practical measuring instruments will remain a dead letter; on the other hand, a definition that may not be quite rigorous theoretically may prove to be of great practical usefulness if the corresponding measurements are simple and can readily be understood by the average operating engineer."

Today's engineers experience better conditions than Karapetoff's generation: available are accurate and versatile instruments capable of measuring any conventional electric quantity defined by the most involving mathematical expressions. The theory of electric circuits, the mechanisms of energy conversion, and the phenomena that govern the flow of electric energy are better understood. The practical experience gained during a century of designing, prototyping, maintaining, and improving the infrastructure needed to sustain the continuous supply of electric energy has created a generation of "average practicing engineers" more competent and better documented than ever before. The need for more efficient equipment, the proliferation of distributed generators, and the deregulation of the electric energy industry, are all forces that drive the quest for a resolution of S that, hopefully, will soon be universally accepted.

Many modern researchers like to emphasize that the apparent power is larger than, or equal to, the active power $S \geq P$, by invoking Schwartz's inequality [2]:

$$\left[\int_a^b f(x)g(x)\, dx \right]^2 \leq \int_a^b [f(x)]^2\, dx \int_a^b [g(x)]^2\, dx$$

This inequality can be readily applied to instantaneous electrical quantities

$$\left[\int_0^T v(t)i(t)\, dt \right]^2 \leq \int_0^T [v(t)]^2\, dt \int_0^T [i(t)]^2\, dt$$

and keeping in mind that the active power squared is

$$P^2 = \left[\frac{1}{T} \int_0^T v(t)i(t)\, dt \right]^2$$

and the squared rms voltage and current are

$$V^2 = \frac{1}{T} \int_0^T [v(t)]^2\, dt \quad \text{and} \quad I^2 = \frac{1}{T} \int_0^T [i(t)]^2\, dt$$

results that $P \leq S$ or $S^2 = P^2 + N^2$, where N is a total nonactive power that together with P contributes to line power loss. While the active power P consists of well agreed upon terms P_1 and $P_H = \sum_{h \neq 1} P_h$, the term $N \geq 0$ is formed by a set of subcomponents. This power appears in literature under different names: nonactive, fictitious, and sometimes is still called reactive power. The main goal of this chapter is to present and discuss the best known resolutions of N and its separation in subcomponents. The selection of the presented theories, or models, was influenced by two reasons:

• The proposed approach or method is a recognized milestone theory having a significant historical role that led to better concepts and developments.
• The proposed method is already accepted by a segment of the engineering community, some implementation was started, and the method is documented in approved standards.

4.1 Constantin I. Budeanu's Method

Budeanu was the first scientist to understand the fact that the apparent power in nonsinusoidal systems has more than two components and can be represented in a three-dimensional system. His theory was detailed in his famous 1927 book, *Reactive and Fictive Powers* [3]. His theory can be explained as follows:

The squared rms harmonic currents can be parted in two orthogonal terms:

$$I_h^2 = (I_h \cos \theta_h)^2 + (I_h \sin \theta_h)^2 \tag{4.1}$$

where θ_h is the phase angle between the harmonic voltage phasor $\mathbf{V_h}$ and the harmonic current phasor $\mathbf{I_h}$.

From equation (4.1) results the expression of the apparent power squared:

$$S^2 = V^2 I^2 = \sum_{h=1}^{v} V_h^2 \left[\sum_{h=1}^{v} (I_h \cos \theta_h)^2 \right] + \sum_{h=1}^{v} V_h^2 \left[\sum_{h=1}^{v} (I_h \sin \theta_h)^2 \right] \tag{4.2}$$

Next step is to apply Lagrange's identity [4]:

$$\sum_{h=1}^{v} A_h^2 \sum_{h=1}^{v} B_h^2 = \left(\sum_{h=1}^{v} A_h B_h \right)^2 + \sum_{m=1}^{v-1} \sum_{n=m+1}^{v} (A_m B_n - A_n B_m)^2 \tag{4.3}$$

to (4.2), thus

$$S^2 = \left(\sum_{h=1}^{v} V_h I_h \cos \theta_h \right)^2 + \left(\sum_{h=1}^{v} V_h I_h \sin \theta_h \right)^2$$

$$+ \sum_{m=1}^{v-1} \sum_{n=m+1}^{v} (V_m I_n \cos \theta_n - V_n I_m \cos \theta_m)^2 + \sum_{m=1}^{v-1} \sum_{n=m+1}^{v} (V_m I_n \sin \theta_n - V_n I_m \sin \theta_m)^2$$

$$= \left(\sum_{h=1}^{v} V_h I_h \cos \theta_h \right)^2 + \left(\sum_{h=1}^{v} V_h I_h \sin \theta_h \right)^2$$

$$+ \sum_{m=1}^{v-1} \sum_{n=m+1}^{v} \left[(V_m I_n)^2 + (V_n I_m)^2 - 2 V_m V_n I_m I_n \cos(\theta_m - \theta_n) \right] \tag{4.4}$$

The apparent power is separated into three terms that define a right angle parallelepiped, Fig. 4.1, such that its diagonal is

$$S^2 = P^2 + Q_B^2 + D_B^2 \tag{4.5}$$

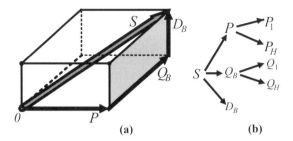

Figure 4.1 Budeanu's apparent power resolution: (a) Three-dimensional representation. (b) Power components.

The first term

$$P = \sum_{h=1}^{\nu} V_h I_h \cos \theta_h \qquad (4.6)$$

is the total active power. The second term

$$Q_B = \sum_{h=1}^{\nu} V_h I_h \sin \theta_h \qquad (4.7)$$

Budeanu called reactive power[1], and the third term

$$D_B = \sqrt{\sum_{m=1}^{\nu-1} \sum_{n=m+1}^{\nu} \left[(V_m I_n)^2 + (V_n I_m)^2 - 2 V_m V_n I_m I_n \cos(\theta_m - \theta_n) \right]} \qquad (4.8)$$

is the *distortion power*.

For years the limits of summations on D_B were wrongly written [5], instead of $\sum_{m=1}^{\nu-1} \sum_{n=m+1}^{\nu}$ one finds in textbooks, research articles, and reports $\sum_m \sum_n$, $\sum_{\substack{m,n=1 \\ m \neq n}}$ or simple $\sum_{m,n}$. The probable reason for this misleading oversight is the fact that not many people compute D_B using equation (4.8). The usual road to the computation of distortion power is through $D_B = \sqrt{S^2 - P^2 - Q_B^2}$.

From (4.8) we find that the distortion power D_B is nil if

$$\theta_m = \theta_n \quad \text{and} \quad V_m I_n = V_n I_m$$

---

[1] The subscript B is used to emphasize that this is Budeanu's definition and to avoid confusion with expressions given in the following sections. The above expressions apply also for situations where interharmonics, subharmonics, and direct current are present. In his original work Budeanu did not specify the limits of the summations.

This means

$$\frac{V_m}{I_m} = \frac{V_n}{I_n} = \cdots \frac{V_h}{I_h} = \cdots = R_e \quad \text{and} \quad \theta_m = \theta_n = \cdots = \theta_h = \cdots$$

We learn from here that only a linear resistance R_e when supplied with nonsinusoidal voltage will operate with $D_B = 0$, $Q_B = 0$, and $S = P$. If the resistance is nonlinear or if it is a function of the frequency (see section 4.8) then $D_B > 0$.

So far Budeanu's model has weathered more than eight decades. Its main significance rests in the fact that by recognizing a nonactive power different than the reactive power he took a first step toward a better understanding of energy flow in circuits with nonlinear loads. Budeanu's powers started to gain AIEE acceptance in the 1930s and were included in the *American Standard Definitions of Electrical Terms* [6] in 1941. Even today, as this book was written, the expressions (4.5), (4.6), and (4.7), and other definitions that resulted directly from (4.4), are still occupying a significant number of pages on *The IEEE Standard Dictionary* [7]. Its past acceptance and popularity among engineers and top scientists is hard to dispute. Modern textbooks authored by highly respected researchers are presenting Budeanu's resolution of apparent power [8,9,10,11] as the right canonical expression.

The truth is that as early as 1935 the physical meaning of the distortion power was questioned by Waldo V. Lyon. He wrote that in his opinion there is a "vital defect" in the way distortion powers add [12]. He supports his claim with the following example: A nonsinusoidal voltage

$$v = \sin(\omega t) + \sin(3\omega t) \quad \text{pu}$$

supplies a branch $R_a + jX_a = 1 + j$ pu, in parallel with a second branch $R_b - jX_b = 1 - j$ pu, as shown in Fig. 4.2. The impedances are in per-unit at fundamental frequency.

For each branch Lyon lists Budeanu's powers as follows:

$$P_a = 0.6 \text{ pu} \qquad P_b = 1.4 \text{ pu}$$
$$Q_a = -0.8 \text{ pu} \qquad Q_b = 0.8 \text{ pu}$$
$$D_a = \sqrt{0.2} \text{ pu} \qquad D_b = \sqrt{0.2} \text{ pu}$$

He writes the following comments: "*These parallel branches are equivalent to a nonreactive resistance at all frequencies. Thus $Q_a + Q_b = 0$. This being true the resulting distortion power*

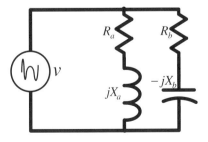

Figure 4.2   W. Lyon's example.

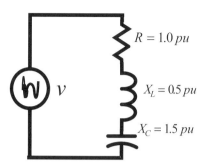

Figure 4.3 Circuit meant to put in evidence the deficiency of Budeanu's method.

calculated at the terminals is zero, whereas the sum of the distortion powers in the two branches is $2\sqrt{0.2}$ *unit. Of what significance then is the sum of the distortion powers in the two branches of a single phase or polyphase network?"*

In 1987 L. S. Czarnecki brought to the attention of the engineering community the fact that Q_B and D_B definitions *"do not possess the attributes which can be related to the power phenomena in circuits with nonsinusoidal waveforms"* [13]. The circuit sketched in Fig. 4.3 helps to explain this fact:

A load with the impedance $1 + 0.5_J - 1.5_J$ pu (measured at fundamental frequency) is supplied with a nonsinusoidal voltage

$$v = \sqrt{2}[\sin(\omega t) + \sin(3\omega t)] \text{ pu}$$

The impedances of this load are $\mathbf{Z_1} = \sqrt{2}\angle - 45°$ pu for the fundamental frequency and $\mathbf{Z_3} = \sqrt{2}\angle 45°$ pu at three times fundamental frequency. The instantaneous current is

$$i = \sin(\omega t + \pi/4) + \sin(3\omega t - \pi/4) \text{ pu}$$

The squared rms voltage and current are $V^2 = 2.0$ pu^2 and $I^2 = 1.0$ pu^2, yielding the powers

$$S = \sqrt{2} \text{ pu} \quad P = 1.0 \text{ pu} \quad Q_B = 0 \quad \text{and} \quad D_B = \sqrt{S^2 - P^2 - Q_B^2} = 1.0 \text{ pu}$$

The result $Q_B = 0$ is not meaningful, it is artificial and it defies the reality of Poynting vector. The instantaneous reactive current is

$$i_q = \frac{1}{\sqrt{2}}[\cos(\omega t) - \cos(3\omega t) \text{ pu}]$$

yielding the instantaneous reactive power

$$p_q = 0.5[\sin(2\omega t) - \sin(6\omega t)] \text{ pu}$$

These two oscillations of power are not fictitious, they are supported by appertaining electromagnetic waves and they coexist, having different frequencies that make their mutual cancellation imposible. Moreover, each one of these power oscillations is sustained by an electromagnetic wave with power densities quantified by means of Poynting vectors.

It may be of historical interest to mention that at a CIGRE meeting in 1928, A. Iliovici, one of the leaders of the French delegation, claimed that "the quantity [distortion power] defined by Mr. Budeanu, probably will never be directly measured. . . " In spite of its obvious deficiencies and of more objections raised in the recent decades by qualified engineers [14,15], the Budeanu method remained quite popular and 60 years later instrumentation manufacturers proved Iliovici wrong [16]. It is regrettable that such an elegant mathematical approach does not have a perfect physical model.

4.2 Stanislaw Fryze's Method

In 1932 Fryze proposed a simple model [17] that had a strong impact on the apparent power concepts developed in the second half of the last century [18,19,20].

Based on the fact that for sinusoidal conditions where the voltage and the current are

$$v = \widehat{V} \sin(\omega t) \quad \text{and} \quad i = \widehat{I} \sin(\omega t - \theta)$$

we separate the current in an active component

$$i_a = \widehat{I} \cos\theta \sin(\omega t) = \widehat{I_a} \sin(\omega t) ; \quad I_a = I \cos\theta = G\,V$$

and a nonactive component, a reactive one

$$i_b = -\widehat{I} \sin\theta \cos(\omega t) = -\widehat{I_b} \sin(\omega t) ; \quad I_b = I \sin\theta = B\,V$$

with G a conductance

$$G = \frac{I \cos\theta}{V} = \frac{V I \cos\theta}{V^2} = \frac{P}{V^2}$$

and B a susceptance

$$B = \frac{I \sin\theta}{V} = \frac{V I \sin\theta}{V^2} = \frac{Q}{V^2}$$

The above concept leads to the well known power triangle, S, P, Q; with $P = V I_a$, $Q = V I_b$ and $S = V\sqrt{I_a^2 + I_b^2}$.

Fryze expanded this approach from sinusoidal to nonsinusoidal situations. The original idea was to consider an instantaneous current component that mimics the voltage waveform, i.e. that is a replica of the voltage waveform. This current i_a, called *instantaneous active current*, is scaled to yield the active power supported by the voltage v and the actual instantaneous current i. Thus the active current is

$$i_a = Gv = G \sum_h \widehat{V_h} \sin(h\omega t + \alpha_h) = \sum_h \widehat{I_{ah}} \sin(h\omega t + \alpha_h) = \sum_h i_{ah} \qquad (4.9)$$

and the active power is

$$P = G \sum_h V_h^2 = GV^2 \tag{4.10}$$

The scaling coefficient G has exactly the same meaning as it had in the sinusoidal case. It is a conductance with the following mathematical properties

$$G = \frac{1}{R} = \frac{i_{a1}}{v_1} = \cdots = \frac{i_{ah}}{v_h} = \cdots = \frac{P}{V^2} = \frac{I_a}{V} \tag{4.11}$$

and

$$G^2 = \frac{I_a^2}{V^2} = \frac{\sum\limits_h i_{ah}^2}{\sum\limits_h v_h^2} = \frac{\sum\limits_h I_{ah}^2}{\sum\limits_h V_h^2} \tag{4.12}$$

The remaining component of the current

$$i_b = i - i_a \tag{4.13}$$

was called by Fryze *wattless current* (in German *Blindstrom*). From the definition of i_a results

$$P = \frac{1}{T} \int_0^T vi \, dt = \frac{1}{T} \int_0^T vi_a \, dt$$

hence the average power of vi_b is nil

$$\frac{1}{T} \int_0^T vi_b \, dt = \frac{1}{T} \int_0^T v(i - i_a) \, dt = 0$$

From this analysis Fryze concluded that the load supplied by the nonsinusoidal voltage can be modeled with the help of a linear conductance G in parallel with a time varying conductance $G'(t) = i_b(t)/v(t) = 1/R'(t)$ or, as interpreted by later generations of engineers, a current source equal to i_b, Fig. 4.4.

It is easy to prove that the three rms currents I, I_a, and I_b form a right-angle triangle; we have

$$I^2 = \frac{1}{T} \int_0^T (i_a + i_b)^2 \, dt = \frac{1}{T} \int_0^T (i_a^2 + i_b^2 + 2i_a i_b) \, dt \tag{4.14}$$

however

$$\frac{1}{T} \int_0^T i_a i_b \, dt = G \frac{1}{T} \int_0^T vi_b \, dt = 0 \tag{4.15}$$

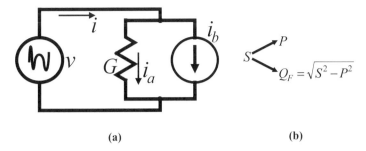

(a)   (b)

Figure 4.4 Fryze's resolution: (a) Equivalent circuit. (b) Components.

thus

$$I^2 = \frac{1}{T} \int_0^T (i_a^2 + i_b^2)\, dt = I_a^2 + I_b^2 \tag{4.16}$$

yielding a result similar to the sinusoidal situation:

$$S^2 = P^2 + Q_F^2 \quad \text{where} \quad S = VI\,; \quad P = VI_a\,; \quad Q_F = VI_b \tag{4.17}$$

with a power factor

$$PF = \frac{P}{S} = \frac{I_a}{I} \tag{4.18}$$

Fryze also introduced the concept of wattless power factor:

$$PF_b = \frac{Q_F}{S} = \frac{I_b}{I}$$

with the property $PF = \sqrt{1 - (PF_b)^2}$.

Fryze's model is mathematically correct, and a direct consequence of the approach used for systems with sinusoidal waveforms, nevertheless, it is only a simplified model, not revealing all the details of the instantaneous power and not representing the true mechanism of energy transfer to the load.

If the actual current has the expression

$$i = \sum_h \widehat{I_h} \sin(h\omega t + \alpha_h - \theta_h) \tag{4.19}$$

the active powers truthful to the Poynting vector reality are

$$P_1 = V_1 I_1 \cos\theta_1 \quad \text{and the components} \quad P_h = V_h I_h \cos\theta_h$$

The actual active powers P_1 and P_h, carried by the Poynting vector, are not represented in Fryze's model. Fryze's approach leads to a set of very different active powers:

$$P_{a1} = G V_1^2 \neq V_1 I_1 \cos\theta_1 = P_1\,; \quad P_{ah} = G V_h^2 \neq V_h I_h \cos\theta_h = P_h$$

It is absolutely correct that $\sum_h P_h = \sum_h P_{ah} = P$, nevertheless $P_1 = V_1 I_1 \cos \theta_1 \neq P_{a1}$ and $P_h \neq P_{ah}$. Moreover, Fryze's model falls short of providing information on the actual 50 or 60 Hz active and reactive power components, P_1 and Q_1, which are the most important power quantities to be measured.

4.3   Manfred Depenbrock's Method

In the 1960s Depenbrock improved on Fryze's method, creating a more detailed model [18,19]. In this case the nonsinusoidal current and voltage components are separated in the fundamental and the total harmonic (section 3.7):

$$i = i_1 + i_H$$

where

$$i_1 = \widehat{I}_1 \sin(\omega t + \alpha_1 - \theta_1) \quad \text{and} \quad i_H = \sum_{h \neq 1} \widehat{I}_h \sin(h\omega t + \alpha_h - \theta_h) \tag{4.20}$$

and

$$v = v_1 + v_H$$

where

$$v_1 = \widehat{V}_1 \sin(\omega t + \alpha_1) \quad \text{and} \quad v_H = \sum_{h \neq 1} \widehat{V}_h \sin(h\omega t + \alpha_h) \tag{4.21}$$

Next, the fundamental current i_1 is separated into two components: i_{p1}, the in-phase, and i_{q1} the in-quadrature component:

$$i = i_{p1} + i_{q1}$$

where

$$i_{p1} = \widehat{I}_1 \cos(\theta_1) \sin(\omega t + \alpha_1) \quad \text{and} \quad i_{q1} = -\widehat{I}_1 \sin(\theta_1) \cos(\omega t + \alpha_1) \tag{4.22}$$

Depenbrock introduced the concept of *conversion current* i_{v1}, defined as the difference between the in-phase current and the fundamental active current, i.e.

$$i_{v1} = i_{p1} - i_{a1} = i_{p1} - Gv_1$$

The rms value of i_{p1}, the in-phase current, is

$$I_{p1} = \frac{P_1}{V_1} = \frac{V_1 I_1 \cos \theta_1}{V_1} = G_1 V_1; \quad G_1 = \frac{I_1 \cos \theta_1}{V_1} = \frac{P_1}{V_1^2} \tag{4.23}$$

Note that $G_1 \neq G = P/V^2$.

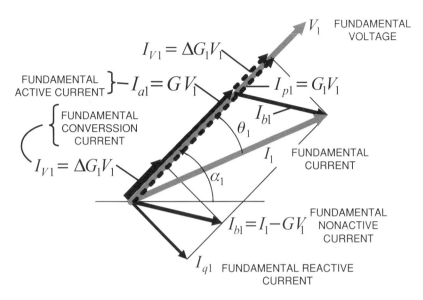

Figure 4.5 Depenbrock's separation of currents: Phasor diagram for the fundamental voltage and currents.

In Fig. 4.5 are shown phasor diagrams of the fundamental voltage and currents. The input data consists of three phasors: $\mathbf{V}_1 = V_1 \angle \alpha_1$, $\mathbf{I}_1 = I_1 \angle(\alpha_1 - \theta_1)$ and $\mathbf{I}_{a1} = GV_1\angle\alpha_1$. The fundamental fictitious (or wattless) current phasor is $\mathbf{I}_{b1} = \mathbf{I}_1 - G\mathbf{V}_1$ and has two components: The fundamental conversion phasor

$$\mathbf{I}_{v1} = \mathbf{I}_{p1} - G\mathbf{V}_1 = (G_1 - G)\mathbf{V}_1 = \Delta G_1 \mathbf{V}_1 \qquad (4.24)$$

where $I_{p1} = I_1 \cos(\theta_1)$ and the fundamental rms reactive current $I_{q1} = I_1 \sin(\theta_1)$.

A set of instantaneous currents correspond to the phasors shown in Fig. 4.5: The fundamental instantaneous current,

$$i_1 = i_{a1} + i_{b1}$$

the fundamental instantaneous conversion current,

$$i_{v1} = \Delta G_1 v_1 \qquad (4.25)$$

and the fundamental instantaneous wattless current

$$i_{b1} = i_{v1} + i_{q1} \qquad (4.26)$$

Depenbrock, rightfully, emphasizes the significance of i_{q1} He writes: "*In practice it is often of interest to know the fundamental portion of the non-active current i_b* [in original i_F] *which is lagging the fundamental of voltage v_1* [in original v_g] *by 90°*.... This current i_{q1} [in original i_Q]

can be found in the same way as the active current i_a [in original i_p] if the reactive power Q_1 is defined to be the average value of the power which comes from the [expression]^2:"

$$Q_1 = \frac{1}{T} \int_0^T v_1 \left(t - \frac{T}{4} \right) i_1 \, dt$$

The harmonic active power $P_H = \sum_{h \neq 1} V_h I_h \cos \theta_h$ is viewed as being caused by the contribution of a *collective active harmonic current* $G_H v_H$ where $G_H = P_H / V_H^2$.

Similarly to the way we separated the fundamental conversion current we find the *instantaneous harmonic conversion current*

$$i_{vH} = G_H v_H - G v_H = (G_H - G) v_H = \Delta G_H v_H \tag{4.27}$$

with an rms value $I_{vH} = |\Delta G_H| V_H$.

The *total conversion current* is $i_v = i_{v1} + i_{vH}$ with the rms value $I_V = \sqrt{I_{v1}^2 + I_{vH}^2}$.

Depenbrock went one step further to extract a *distortion current* $i_D = i_b - i_{q1} = i - i_a - i_{q1}$, this is a current associated with a nonactive power that was separated from the reactive current. The distortion current consists of the total conversion current i_v and what is left from the harmonic current after subtracting the collective active harmonic current $G_H v_H$, i.e. $i_D = i_v + i_H - G_H v_H$.

In Fig. 4.6 is sketched Depenbrock's chart of instantaneous current flow: The line current is divided according to Fryze

$$i = i_a + i_b = G(v_1 + v_H) + i_b \tag{4.28}$$

also from the right end of the flow chart we find

$$i = i_1 + i_H = i_{p1} + i_{q1} + i_H \tag{4.29}$$

The wattless current i_b can be expressed in a few ways:

$$i_b = i - i_a = (i_{p1} + i_{q1} + i_H) - (G v_1 + G v_H) = i_{v1} + i_{q1} + i_H - G v_H \tag{4.30}$$

A residual instantaneous current $i_N = i_H - G_H v_H$ helps to complete the flow chart

$$i_b = i_{v1} + i_{q1} + i_{vH} + i_N \tag{4.31}$$

This is exactly Kirchhoff's current law at the cross-section X–X in Fig. 4.6. The distortion current

$$i_D = i_b - i_{q1} = i_{v1} + i_{vH} + i_N = i_v + i_N \tag{4.32}$$

enters the cross-section XX–XX.

---

^{2} This is the fictitious average power produced by the voltage v_1 shifted $90°$ and the current i_1 (see Appendix VIII).

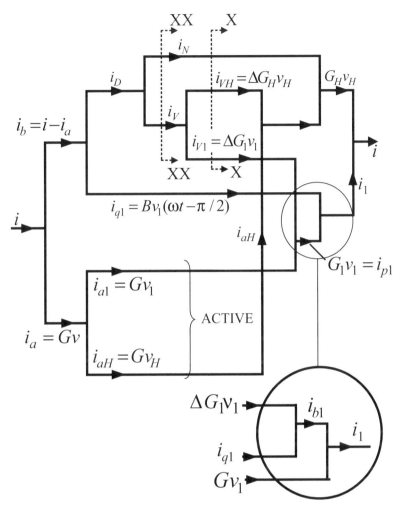

Figure 4.6 Depenbrock's approach: Current separation in elementary components for a single-phase system. Source: M. Depenbrock, "The FBD-Method, A Generally Applicable tool for Analyzing Power Relations," IEEE Trans. On Power Systems, Vol.8, No.2, May 1993, pp.381–87

An equivalent susceptance B ties the rms fundamental voltage to the fundamental reactive power, $|Q_1| = V_1 I_{q1}$, hence $B = Q_1/V_1^2$. Depenbrock defines two reactive powers; one is the fundamental reactive power Q_1, the second is a *complimentary reactive power* $Q_c = V_H I_{q1}$. This grouping can be traced to the definition

$$Q^2 = (V I_{q1})^2 = (V_1^2 + V_H^2)I_{q1}^2 = Q_1^2 + Q_c^2$$

All six instantaneous currents i_{a1}, i_{aH}, i_{q1}, i_{v1}, i_{vH} and i_N are mutually orthogonal (demonstrated in the same way as for (4.16)). Based on this property Depenbrock presented the following apparent power resolution:

$$S = VI = V\sqrt{I_{a1}^2 + I_{aH}^2 + I_{q1}^2 + I_{v1}^2 + I_{vH}^2 + I_N^2}$$

$$S^2 = P^2 + Q_F^2; \quad P = VI_a = V\sqrt{I_{a1}^2 + I_{aH}^2}; \quad Q_F = VI_b$$

$$Q_F^2 = D_D^2 + Q^2; \quad D_D = VI_D$$

$$D_D^2 = D_C^2 + D_R^2; \quad D_C = VI_V = V\sqrt{I_{v1}^2 + I_{vH}^2}; \quad D_R = VI_N$$

$$Q^2 = Q_1^2 + Q_c^2; \quad Q = VI_{q1} = \sqrt{V_1^2 + V_H^2}\,I_{q1}$$

$$S = \sqrt{P^2 + D_C^2 + D_R^2 + Q_1^2 + Q_c^2} \tag{4.33}$$

The complete equivalent circuit for Depenbrock's method is given in Fig. 4.7. The circuit consists of a set of conductances G, ΔG_1, ΔG_H and a susceptance B, each in series with a voltage v_1 or v_H. The purpose of the series voltage source is to provide the needed voltage across the respective admittance. The upper rail, that carries current i, has the potential $v_1 + v_H$ with respect to the reference point. If, for example, the fundamental active power branch carries the current

$$i_{a1} = Gv_1 = G(v - v_H)$$

it is necessary to insert in series with G the voltage source $-v_H$ in order to obtain the desired current:

$$i_{a1} = G(v_1 + v_H - v_H) = Gv_1$$

This very involving model covers a set of essential components of S, and when it was first presented it meant a significant step forward. It does not provide only components separation in active and nonactive powers, it also reveals the nature of nonactive components, thus suggesting that adequate compensation systems, such as active, passive, and hybrid filters can be designed and controlled to cancel partially, or totally, the nonactive currents.

The major drawback of this approach stems from the fact that, just like Fryze's method, it pivots around the active power $P_{a1} = GV_1^2 = VI_{a1}$ instead of $P_1 = V_1 I_1 \cos\theta_1$, the true fundamental active power. Moreover, the complementary reactive power Q_c seems to be only a mathematical artifice without a solid physical base. Depenbrock's apparent power symbolic resolution is presented in Fig. 4.8

As it will be shown in Chapter 6, Depenbrock's method has had a major influence on the way polyphase systems are analyzed by many researchers.

4.4   Leszek Czarnecki's Method

This approach, presented in 1984 [20], was meant to improve on the limitations of Fryze's model, namely to provide more information on the nature of the load and the type of compensator needed to ameliorate the load's power factor. Czarnecki's equivalent circuit is shown

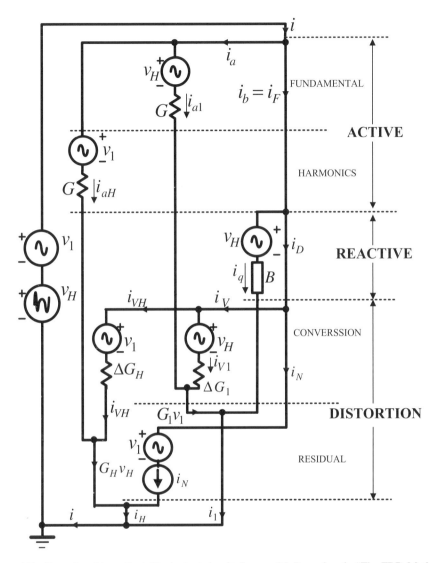

Figure 4.7 Depenbrock's method: Equivalent circuit. Source: M. Depenbrock, "The FBD-Method, A Generally Applicable tool for Analyzing Power Relations," IEEE Trans. On Power Systems, Vol.8, No.2, May 1993, pp.381–87

in Fig. 4.9a. The current source i_b in Fig. 4.4, was replaced with two current sources: i_r, the instantaneous reactive current, and i_s, the *instantaneous scattered current*.

The conductance $G_e = P/V^2$ and the current have exactly the same meaning as in Fryze's and Depenbrock's approaches (4.11, 4.23); thus if the input voltage is

$$v = \sum_{h \in M} \widehat{V}_h \cos(h\omega t + \alpha_h) \tag{4.34}$$

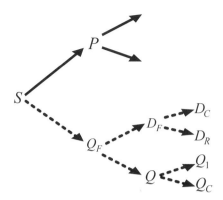

Figure 4.8   Depenbrock's method: Power components.

the instantaneous active current i_a has a waveform that replicates the voltage waveform

$$i_a = G_e v = G_e \sum_{h \in M} \widehat{V}_h \cos(h\omega t + \alpha_h) \tag{4.35}$$

where M represents the complete set of harmonics and interharmonics that constitute the instantaneous voltage v.

The nature of the currents i_s and i_r is explained with the help of the phasor diagram given in Fig. 4.9b.

The phasor $\mathbf{I}_{ah} = G_e \mathbf{V}_h$ is the active harmonic current of order h, it is in-phase with the phasor $\mathbf{V}_h$ and is not necessarily equal to $\mathbf{I}_{ph} = G_h \mathbf{V}_h = I_h \cos \theta_h$.

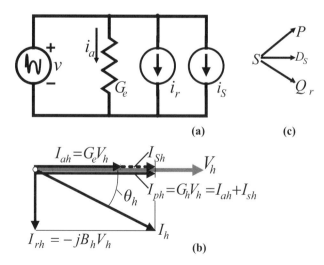

Figure 4.9   Czarnecki's method: (a) Equivalent circuit. (b) Phasor diagram for the h-order harmonics. (c) Components.

Similar to Depenbrock's conversion currents, (4.24) and (4.27), a scattered current harmonic phasor of order h in-phase with the harmonic voltage v_h is defined:

$$I_{sh} = I_{ph} - I_{ah} = (G_h - G_e)V_h = \Delta G_{eh} V_h \tag{4.36}$$

The remaining component of I_h is the in-quadrature phasor

$$\mathbf{I}_{rh} = -J B_h \mathbf{V}_h \tag{4.37}$$

where B_h is a susceptance. The instantaneous value of i_{rh} is

$$i_{rh} = \Im m \left\{ -B_h \widehat{V}_h \exp[J(h\omega t + \alpha_h)] \right\} = \Re e \left\{ J B_h \widehat{V}_h \exp[J(h\omega t + \alpha_h)] \right\}$$
$$= B_h \widehat{V}_h \sin(h\omega t + \alpha_h) \tag{4.38}$$

In the same way we can express i_{sh}

$$i_{sh} = \Re e \left\{ (G_h - G_e)\widehat{V}_h \exp[J(h\omega t + \alpha_h)] \right\} = (G_h - G_e)\widehat{V}_h \cos(h\omega t + \alpha_h) \tag{4.39}$$

The instantaneous value of the total scattered current has the expression

$$i_s = \Re e \left\{ \sum_{h \in M} (G_h - G_e)\widehat{V}_h \exp[J(h\omega t + \alpha_h)] \right\} \tag{4.40}$$

and the instantaneous value of the total reactive current is

$$i_r = \Re e \left\{ \sum_{h \in M} J B_h \widehat{V}_h \exp[J(h\omega t + \alpha_h)] \right\} \tag{4.41}$$

The three currents i_a, i_s and i_r are mutually orthogonal, hence

$$I^2 = I_a^2 + I_s^2 + I_r^2 \tag{4.42}$$

where

$$I_a = \frac{P}{V} = G_e V = G_e \sqrt{\sum_{h \in M} V_h^2} \tag{4.43}$$

$$I_s = \sqrt{\sum_{h \in M} (G_h - G_e)^2 V_h^2} \tag{4.44}$$

and

$$I_r = \sqrt{\sum_{h \in M} (B_h V_h)^2} \tag{4.45}$$

Equation (4.42) leads to apparent power resolution in three components, Fig. 4.9c:

$$S = \sqrt{P^2 + D_s^2 + Q_r^2} \tag{4.46}$$

where

$$P = V I_a = G_e V^2 \tag{4.47}$$

is the active power,

$$D_s = V I_s = V \sqrt{\sum_{h \in M} (G_h - G_e)^2 V_h^2} \tag{4.48}$$

was named by Czarnecki *scattered power* and the last term

$$Q_r = V I_r = V \sqrt{\sum_{h \in M} (B_h V_h)^2} \tag{4.49}$$

is the [*collective*] *reactive power*[3].

We see here a model definitely superior to Fryze's, a model that, unlike Budeanu's, has a collective reactive power with subcomponents that cannot cancel one another out, $(B_h V_h)^2 > 0$, and are mathematical truthfully to the actual oscillations of energy.

The scattered power D_s has no precedent in engineering literature. Its instantaneous correspondent $v i_s$ is composed of active instantaneous powers, each of a different frequency, each elementary component having an average value, but the overall sum of the average powers is nil. This scattered power is definitely an ingenious mathematical definition, nevertheless the electromagnetic field theory does not provide evidence or support for such components.

The major drawback of this method, and of all the methods that focus on the active current i_a as the key component, is that the most important power, the fundamental active power, P_1, is not a salient, easily identified component of S. As a matter of fact all models based on Fryze's approach do not use or advocate the need to measure the actual fundamental active power $P_1 = V_1 I_1 \cos \theta_1$.

4.5 The Author's Method

The power frequency (60/50 Hz or fundamental) apparent, active, and reactive powers are the essential components among all the components of the apparent power. The electric energy is generated with nearly pure sinusoidal voltage and currents and the end-users, who buy the electric energy, expect a high quality product, i.e. the provider of electric energy is expected to deliver reasonable sinusoidal voltage waveforms that support the useful energy $P_1 t$. The harmonic powers P_h are often considered electromagnetic pollution—a by-product of the energy conversion process that takes place within the nonlinear loads.

A distribution system cannot perform without reactive power. The useful, fundamental magnetizing flux in transformers and ac motors is supported by the fundamental reactive

---

[3] The IEEE Std. 1459–2010 recommends the term "Reactive Power" to be reserved for the fundamental reactive power only.

current. The voltage profile along a distribution feeder is tied to the fundamental reactive power, and the power loss in the feeder is also dependent on the amount of reactive power flowing through the feeder.

Thus, it makes good sense to separate P_1 and Q_1 from the rest of the powers [21]. This can be easily obtained by starting from the separation of rms currents and voltages according to the conventional grouping used in (3.68) and (3.69) where

$$V^2 = V_1^2 + V_H^2 \; ; \quad V_H^2 = \sum_{h \neq 1} V_h^2 \qquad (4.50)$$

and

$$I^2 = I_1^2 + I_H^2 \; ; \quad I_H^2 = \sum_{h \neq 1} I_h^2 \qquad (4.51)$$

From (4.50) and (4.51) the result is that the apparent power squared has four terms [21]

$$
\begin{aligned}
S^2 = V^2 I^2 &= \left(V_1^2 + V_H^2 \right)\left(I_1^2 + I_H^2 \right) \\
&= (V_1 I_1)^2 + (V_1 I_H)^2 + (V_H I_1)^2 + (V_H I_H)^2 = S_1^2 + D_I^2 + D_V^2 + S_H^2
\end{aligned}
\qquad (4.52)
$$

The first term is the *fundamental or 60/50 Hz apparent power*

$$S_1 = (V_1 I_1) = \sqrt{P_1^2 + Q_1^2} \quad \text{(VA)} \qquad (4.53)$$

The remaining three terms make the *nonfundamental (non-60/50 Hz) apparent power*

$$S_N = \sqrt{D_I^2 + D_V^2 + S_H^2} \quad \text{(VA)} \qquad (4.54)$$

where

$$D_I = V_1 I_H = V_1 \sqrt{\sum_{h \neq 1} I_h^2} \quad \text{(var)} \qquad (4.55)$$

is the *current distortion power*. This nonactive power gives useful information on the amount of VA tied to the current distortion and usually it is the dominant term of S_N, (4.54).

$$D_V = V_H I_1 = I_1 \sqrt{\sum_{h \neq 1} V_h^2} \quad \text{(var)} \qquad (4.56)$$

is the *voltage distortion power*, and is proportional to the fundamental component of the current, I_1 and the total harmonic voltage V_H. It reveals the amount of volt–amper–reactives caused by voltage distortion.

The last term is the *harmonic apparent power*

$$S_H = V_H I_H = \sqrt{\sum_{h \neq 1} V_h^2 \sum_{h \neq 1} I_h^2} \quad \text{(VA)} \tag{4.57}$$

Being significantly smaller than the other components S_H makes the least contribution to S_N. As determined in section 3.7, S_H contains the active harmonic power P_H,

$$S_H = \sqrt{P_H^2 + D_H^2} \tag{4.58}$$

where

$$P_H = \sum_{h \neq 1} P_h = \sum_{h \neq 1} V_h I_h \cos \theta_h$$

and

$$D_H = \sqrt{\left(\sum_{h \neq 1}^{v} V_h I_h \sin \theta_h\right)^2 + \sum_{m=1}^{v-1} \sum_{n=m+1}^{v} [(V_m I_n)^2 + (V_n I_m)^2 - 2 V_m V_n I_m I_n \cos(\theta_m - \theta_n)]}$$

$$\tag{4.59}$$

is the *harmonic distortion power*.

Keeping in mind that S_1 is the key component—a power that plays the major role in power flow studies, the term that covers the fundamental active and reactive powers P_1 and Q_1—it makes good sense to normalize the other powers using S_1 as base power. Starting from the definitions of the total harmonic voltage and current distortions

$$THD_I = \frac{I_H}{I_1} = \sqrt{\sum_{h \neq 1} \frac{I_h^2}{I_1^2}} = \sqrt{\left(\frac{I}{I_1}\right)^2 - 1}$$

and

$$THD_V = \frac{V_H}{V_1} = \sqrt{\sum_{h \neq 1} \frac{V_h^2}{V_1^2}} = \sqrt{\left(\frac{V}{V_1}\right)^2 - 1}$$

we find

$$D_I = V_1 I_H = S1 \frac{V_1 I_H}{V_1 I_1} = S_1(THD_I) \tag{4.60}$$

$$D_V = V_H I_1 = S1 \frac{V_H I_1}{V_1 I_1} = S_1(THD_V) \tag{4.61}$$

$$S_H = V_H I_H = S1 \frac{V_H I_H}{V_1 I_1} = S_1(THD_V)(THD_I) \tag{4.62}$$

hence

$$S_N = S_1 \sqrt{THD_I^2 + THD_V^2 + [(THD_V)(THD_I)]^2} \qquad (4.63)$$

and

$$S^2 = S_1^2 \{1 + THD_I^2 + THD_V^2 + [(THD_V)(THD_I)]^2\} \qquad (4.64)$$

The last three expressions reflect the impact of current and voltage distortions on the total apparent power S. Since typically $0.01 < THD_V < 0.07$ and $0.05 < THD_I < 1.2$ the voltage distortion power D_V is less significant than D_I. In Fig. 4.10 are shown the graphs S_N/S_1 in function of THD_I (in %) when THD_V is parameter. We observe that for $THD_I > 20\%$ and $THD_V < 5\%$ the trend is

$$S_N \approx S_1(THD_I) \qquad (4.65)$$

The error for this approximation is graphed in Fig. 4.11a. A significantly better approximation that leads to acceptable errors is

$$S_N \approx S_1 \sqrt{THD_I^2 + THD_V^2} \qquad (4.66)$$

For $THD_V \le 6\%$, the error ϵ caused by (4.66), Fig. 4.11b, is less than 0.20% for any value of THD_I.

The resolution of S is sketched in Fig. 4.12. If we return to Fig. 3.9c we observe that all five instantaneous powers, $v_1 i_{p1}$, $v_1 i_{q1}$, $v_1 i_H$, $v_H i_1$ and $v_H i_H$, are materialized in the resolution of

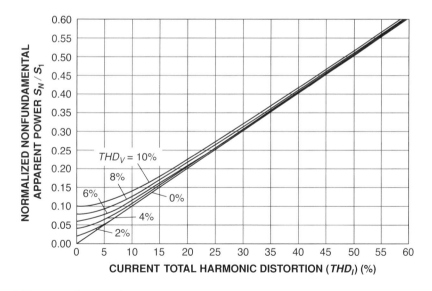

Figure 4.10   Normalized nonfundamental apparent power, S_N/S_1, versus THD_I, when THD_V is parameter [15]. Source: IEEE Working Group on Nonsinusoidal Situations (A. Emanuel Chairman), "Practical Definitions for Powers in Systems with Nonsinusoidal Waveforms...," IEEE Trans. On Power Delivery, Vol. 11, No. 1, January 1996, pp. 79–101

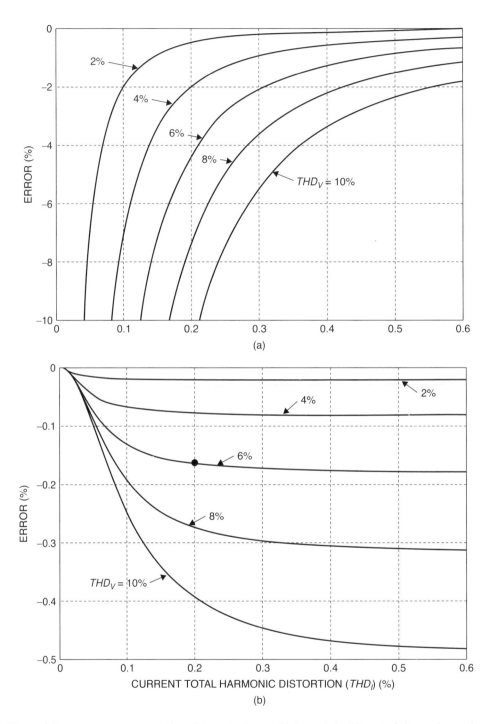

Figure 4.11 Error in the computation of S_N: (a) When (4.65) is used. (b) When S_H is ignored, equation (4.66).

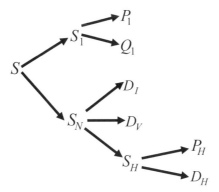

Figure 4.12   Apparent power resolution according to IEEE Std. 1459–2010. [21,14,22].

S through P_1, Q_1, D_I, D_V and S_H, respectively. The amplitudes that quantify the flow of these powers are found to be well correlated with the five basic components of S as revealed by the Poynting vector studies.

This method was adopted in the IEEE Std. 1459–2010, Definitions for the Measurement of Electric Power Quantities [22]. This standard reserves the reactive power name only for Q_1 and for $Q_h = V_h I_h \sin(\theta_h)$. The reactive power belongs to the group of nonactive powers that includes D_I, D_V, and D_H. Today's varmeters, when connected in circuits with nonsinusoidal waveforms, yield readings that are not equal to any of the four nonactive powers (see Appendix VIII).

4.6   Comparison Among the Methods

A fair way to compare the merits and shortcomings of the models previously presented is by using a numerical example; we start from a set of voltage and current measurements implemented at the terminals of a single-phase load with the following instantaneous voltage and current:

$$v = \sqrt{2}[100\sin(\omega t) + 15\sin(3\omega t + 10°) + 20\sin(7\omega t + 110°)]$$
$$i = \sqrt{2}[60\sin(\omega t - 30°) + 60\sin(3\omega t + 105°) + 20\sin(7\omega t + 204°)]$$

The rms values of the voltage and current harmonics are:

$$V_1 = 100 \text{ V} ; \quad V_3 = 15 \text{ V} ; \quad V_5 = 20 \text{ V}$$
$$I_1 = 60 \text{ A} ; \quad I_3 = 60 \text{ A} ; \quad I_5 = 30 \text{ A}$$

with the current/voltage phase angles:

$$\theta_1 = 30° ; \quad \theta_3 = -95° ; \quad \theta_5 = -94°$$

The rms values of the voltage and current are

$$V = \sqrt{V_1^2 + V_3^2 + V_5^2} = 103.08 \text{ V} \quad \text{and} \quad I = \sqrt{I_1^2 + I_3^2 + I_5^2} = 90.0 \text{ A}$$

yielding the apparent power

$$S = VI = 9277 \text{ VA}$$

the fundamental active power

$$P_1 = V_1 I_1 \cos(\theta_1) = 5196 \text{ W}$$

the harmonic active power

$$P_H = V_3 I_3 \cos(\theta_3) + V_5 I_5 \cos(\theta_5) = -120.29 \text{ W}$$

the fundamental reactive power

$$Q_1 = V_1 I_1 \sin(\theta_1) = 3000 \text{ var}$$

and the harmonic reactive powers

$$Q_3 = V_3 I_3 \sin(\theta_3) = -896.57 \text{ var} \quad \text{and} \quad Q_5 = V_5 I_5 \sin(\theta_5) = -598.54 \text{ var}$$

a) **Budeanu**'s model gives:

Table 4.1 Powers Resulting
from Budeanu's Model

S	(VA)	9277
P	(W)	5076
Q_B	(var)	1505
D_B	(var)	7618

One immediately observes that the total reactive power $Q_B = Q_1 + Q_3 + Q_5 = 1505$ var $< Q_1 = 3000$ var. This result points to a significant deficiency of the model.

Since D_B is calculated from the values of S, P, and Q_B, it results that also D_B is a misleading result.

b) **Fryze**'s model yields a simpler set of results:

Table 4.2 Powers Resulting
from Fryze's Model

S	(VA)	9277
P	(W)	5076
Q_F	(var)	7765

This table hides some revealing data; the conductance $G = P/V^2 = 0.478$ S. The rms total active current $I_a = P/V = 49.24$ A, the components of the active current are

$$I_{a1} = 47.77 \text{ A} ; \quad I_{a3} = 7.17 \text{ A} ; \quad I_{a5} = 9.55 \text{ A}$$

leading to the corresponding active powers

$$P_{a1} = 4777 \text{ W} ; \quad P_{a3} = 107.49 \text{ W} ; \quad P_{a5} = 191.09 \text{ W}$$

The wattless current is

$$I_b = \sqrt{I^2 - I_a^2} = 75.33 \text{ A}$$

Except that $\sum_h P_{ah} = \sum_h P_h$, the individual active powers P_{a1}, P_{a3}, and P_{a5} lack any correlation to the true active powers, $P_1 = 5196$ W, P_3 and P_5, $(P_3 + P_5 = -120.29$ W). Since the studied load is a polluter, the harmonic active powers P_3 and P_5 are negative. This very important fact is hidden by Fryze's result and by other methods based on Fryze's model.

c) **Depenbrock**'s concept is a more complete and comprehensive approach. The instantaneous power flow studies presented in Chapter 3 are quite similar to Depenbrock's approach. The main difference is the fact that in Chapter 3 the powers are separated according to their strict electromagnetic nature and the power $P_1 = V_1 I_1 \cos\theta_1$ is singled as a main component that should be separated from P_H. In Depenbrock's model, as in Fryze's, the instantaneous current i is divided in active i_a and fictitious (wattless) i_b, the rms fundamental active current being GV_1 with the fundamental active power GV_1^2 instead of P_1.
The distortion current is

$$I_D = \sqrt{I_b^2 - I_{q1}^2} = 69.10 \text{ A}$$

To determine the conversion currents we first need the conductances

$$G_1 = P_1/V^2 = 0.489 \text{ S} \quad \text{and} \quad G_H = P_H/V_H^2 = -0.192 \text{ S}$$

and

$$\Delta G_1 = G_1 - G = 0.011 \text{ S} \quad \text{and} \quad \Delta G_H = G_H - G = -0.67 \text{ S}$$

this leading to the conversion current components

$$I_{v1} = |\Delta G_1| V_1 = 1.132 \text{ A} \quad \text{and} \quad I_{vH} = |\Delta G_H| V_H = 16.75 \text{ A}$$

and

$$I_v = \sqrt{I_{v1}^2 + I_{vH}^2} = 16.79 \text{ A}$$

enabling the computation of the residual current

$$I_N = \sqrt{I_D^2 + I_v^2} = 67.03 \text{ A}$$

The reactive powers are

$$Q = V I_{q1} = 3092 \text{ var} \quad \text{and} \quad Q_1 = V_1 I_{q1} = 3000 \text{ var}$$

thus

$$Q_c = \sqrt{Q^2 - Q_1^2} = 750 \text{ var}$$

The corresponding powers are summarized in the following table:

Table 4.3 Powers Resulting from Depenbrock's Model

S	(VA)	9277
P	(W)	5076
P_{a1}	(W)	4777
P_{aH}	(W)	299
D_C	(var)	1731
D_R	(var)	6909
Q_1	(var)	3000
Q_c	(var)	750

The separation of Q_1 is a definite asset of this method, however the quantity Q_c has an artificial role and so are the quantities P_{a1} and P_{aH}. Both the nonactive powers D_r and D_C provide good information since both can be compensated by means of hybrid and active filters.

d) Czarnecki's model stems straight from Fryze's without having the complexity of Depenbrock's or his strict adherence to the circuit theory. The main innovation is the scattered current. In this example the rms components of the scattered current are

$$I_{s1} = I_1 \cos(\theta_1) - I_{a1} = 4.19 \text{ A} ; \quad I_{s3} = I_3 \cos(\theta_3) - I_{a3} = -12.395 \text{ A}$$
$$I_{s5} = I_5 \cos(\theta_5) - I_{a5} = -11.65 \text{ A}$$

giving a total rms scattered current

$$I_s = \sqrt{I_{s1}^2 + I_{s3}^2 + I_{s5}^2} = 17.52 \text{ A}$$

The reactive current has the components

$$I_{q1} = Q_1/V_1 = 30.0 \text{ A} ; \quad I_{q3} = Q_3/V_3 = -59.77 \text{ A} ; \quad I_{q5} = Q_5/V_5 = -29.93 \text{ A}$$

giving a total reactive current

$$I_q = \sqrt{I_{q1}^2 + I_{q3}^2 + I_{q5}^2} = 73.27 \text{ A}$$

The results are summarized in the following table:

Table 4.4 Powers Resulting from Czarnecki's Model

S	(VA)	9277
P	(W)	5076
D_s	(var)	1806
Q_r	(var)	7552

The reactive power Q_r helps size the bulk of the power that oscillates between the source and the load. Its main component Q_1 provides information on the size of a static compensator dedicated to minimizing the fundamental rms current. The usefulness of D_s, however, is harder to fathom; the elementary scattered powers are[4]:

$$P_{s1} = V_1 I_{s1} = 418.87 \text{ W} ; \quad P_{s3} = V_3 I_{s3} = -185.93 \text{ W} ; \quad P_{s5} = V_5 I_{s5} = -232.94 \text{ W}$$

The summation $P_{s1} + P_{s3} + P_{s5} = 0$. These scattered powers cannot be compensated by means of tuned filters, and require active devices able to inject an instantaneous compensation current $i_{comp} = -i_s$.

In practice dynamic and hybrid compensators are designed to compensate a part of i_{q1} and as much of i_H. In section 3.7 it was emphasized that the physical nature of the nonactive powers with single oscillations such as p_{q1}, p_{qh}, with Q_1 and Q_h amplitudes, respectively, is not at all different from the physical nature of the powers with double oscillations, such as p_{DI}, p_{DV}, and $p_{m,n}$, with amplitudes $V_m I_n$. This observation leads to the conclusion that nonactive powers with single or double oscillations are compensated in the same way.

e) Emanuel's method uses the total harmonic distortions to compute the nonactive powers. In this case

$$THD_V = \sqrt{\left(V_3^2 + V_5^2 \right) / V_1^2} = 0.25 \qquad THD_I = \sqrt{\left(I_3^2 + I_5^2 \right) / I_1^2} = 1.118$$

The fundamental apparent power is

$$S_1 = V_1 I_1 = 6000 \text{ VA}$$

---

[4] The scattered power D_s is measured in (var) since it is a nonactive power, but each elementary component has an average power measured in Watt.

and the nonfundamental apparent power

$$S_N = \sqrt{S^2 - S_1^2} = 7075 \text{ VA}$$

The S_N components are computed from (4.60), (4.61), and (4.62). All the powers are listed in the following table:

Table 4.5 Powers Resulting from Emanuel's Model

S	(VA)	9277
S_N	(VA)	7075
S_1	(VA)	6000
S_H	(VA)	1677
$P1$	(W)	5196
P_H	(W)	-120.29
Q_1	(var)	3000
D_I	(var)	6708
D_V	(var)	1500
D_H	(var)	1677

The useful power is the fundamental active power P_1. The fundamental apparent power S_1 and the active power P_1 yield the displacement power factor, or fundamental power factor $PF_1 = P_1/S_1 = 0.866$. The nonfundamental apparent power S_N is a fair estimate of the size of the harmonics compensator, while the fundamental reactive power Q_1 shades light on the amount of kvar needed to improve PF_1.

The measurement of the harmonic active power $P_H = -120.29$ W is a rough indicator revealing that the load is polluting. However, it is the current distortion power D_I that quantifies the amount of harmonic pollution caused by the load, or to the load. The voltage distortion power D_V is due to voltage distortion that may be caused by more than one consumer. It is the responsibility of the electric energy provider to oversee the voltage quality and reduce the THD_V within the limits recommended by standards or agreed with the consumers.

4.7   Power Factor Compensation

In circuits with nonsinusoidal waveforms, the load power factor PF has exactly the same definition and meaning as for sinusoidal conditions: it is the ratio between the actual active power supplied to the load and the maximum power that could be supplied to the receiving end of the line, pending two conditions:

- The line power loss remains unchanged. If the skin effect is negligible this condition is equivalent to keeping the rms current unchanged.
- The load voltage is maintained with the same rms value and waveform.

The second requirement has started to be debatable and may need some reconsideration [23]. If the purpose of the considered load is to convert electric energy into heat, then the

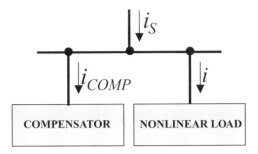

Figure 4.13   Power factor compensation: Concept.

constant rms voltage criterion is perfectly valid. If the energy is converted by the load in mechanical energy, then it is important to maintain the same mechanical output, i.e. torque or force and velocity. If a battery is charged or a dc motor driven, then the direct voltage applied to the dc load is the quantity that should be maintained unchanged. Depending on the type of rectifier and the way the dc is filtered, the direct voltage value may be dependent on the peak input voltage or the mean of the absolute voltage. Thus the definition of S may be more complicated[5] than what is conventionally accepted, especially since in practice there are always many types of loads monitored by one meter.

Due to a lack of further information on this topic, the following explanations will adhere to the traditional approach and assume the rms voltage invariance as a satisfactory condition for single-phase systems.

Under nonsinusoidal conditions it is possible to express the power factor in the following way:

$$PF = \frac{P}{S} = \frac{P_1 + P_H}{\sqrt{S_1^2 + S_N^2}} = \frac{1 + \dfrac{P_H}{P_1}}{\sqrt{1 + THD_I^2 + THD_V^2 + (THD_I THD_V)^2}} PF_1 \qquad (4.67)$$

where

$$PF_1 = \cos\theta_1 = \frac{P_1}{S_1}$$

is the *fundamental power factor*, or the 60/50 Hz *power factor*, also known as the *displacement power factor*.

The concept of power factor improvement for nonsinusoidal conditions is sketched in Fig. 4.13 [24]. The compensator current i_{comp} cancels partially, or totally, the nonactive components of the load current i and the active components of the harmonic currents. By compensation it is meant that the compensator acts like a "sink" for the undesired harmonic current components as well as for the fundamental reactive current.

---

[5] This topic is treated in more detail in Chapter 6.

Harmonic, or tuned, filters are a prevalent technique of power factor compensation [10]. The tuned R–L–C branches series resonate at the harmonic of order m, $m\omega L = 1/m\omega C$ providing a low impedance path, $Z_{Fm} = R$ at the frequency $m\omega$. At the same time such a branch provides a compensating reactive power at the fundamental frequency[6]

$$Q_{F1} \approx C\omega V_1^2 \frac{m^2}{m^2 - 1}$$

A few researchers [25,26] have proposed resolutions of S based on the simplest possible power factor compensator, a linear capacitance connected in parallel with the nonlinear load. Such a capacitance is expected to minimize the rms value of the line current i_s. If the load is supplied with the voltage $v = \sum \widehat{V}_h \sin(h\omega t + \alpha_h)$, the load current is separated in the in-phase components

$$i_p = \sum_h \widehat{I}_h \cos\theta_h \sin(h\omega t + \alpha_h) \qquad (4.68)$$

and the in-quadrature components

$$i_q = -\sum_h \widehat{I}_h \sin\theta_h \cos(h\omega t + \alpha_h) \qquad (4.69)$$

A capacitor C connected at the load bus will cause the flow of the current

$$i_{comp} = \sum_h \widehat{V}_h h\omega C \cos(h\omega t + \alpha_h) \qquad (4.70)$$

Comparing (4.69) with (4.70) we find that some terms of i_q may be reduced by i_{comp}. The total reactive current is

$$i_{sq} = i_q + i_{comp} = \sum_h (\widehat{V}_h h\omega C - \widehat{I}_h \sin\theta_h)\cos(h\omega t + \alpha_h) \qquad (4.71)$$

---

[6] The filter's branch impedance at fundamental frequency is

$$\mathbf{Z}_{F1} = R - jX_{F1}$$

where the equivalent reactance is $X_{F1} = (1/\omega C) - \omega L$.
   The key design expression is $L = 1/m^2\omega^2 C$, (meaning that the filter's branch resonates at $m\omega$ rad/s). Substitution of L into the expression of X_{F1} gives

$$X_{F1} = \frac{1}{\omega C} - \frac{1}{m^2\omega C} = \frac{1}{\omega C}\left(1 - \frac{1}{m^2}\right)$$

The fundamental reactive power provided by this tuned branch is

$$Q_{F1} = \frac{V_1^2}{R^2 + X_{F1}^2} X_{F1}$$

Usually $20 < X_{F1}/R < 150$, and $Q_{F1} \approx V_1^2/X_{F1}$.

The squared rms line reactive current is

$$I_{sq}^2 = \sum_h (V_h h\omega C - I_h \sin\theta_h)^2 = \sum_h \left[(V_h h\omega C)^2 + (I_h \sin\theta_h)^2 - 2V_h I_h h\omega C \sin\theta_h \right]$$

(4.72)

The best capacitance, C_{opt} is found from the condition

$$\frac{\partial}{\partial C}(I_{sq}^2) = \sum_h \left[2(V_h h\omega)^2 C - 2V_h I_h h\omega \sin\theta_h \right] = 0$$

that yields

$$\omega C_{opt} = \frac{\sum_h h V_h I_h \sin\theta_h}{\sum_h (h V_h)^2}$$

(4.73)

The same computation can be carried for a tuned filter meant to provide a low impedance path for a dominant harmonic and at the same time reduce the line reactive current (see problem 4.10.)

This type of power factor compensation has a major drawback, namely there is the possibility of resonances at one or more harmonic frequencies with the equivalent components of the Thévenin's circuit supplying the nonlinear load.

A simple numerical example will help to illustrate such a situation. In Fig. 4.14a is shown a nonlinear load represented by a resistance $R = 1414.2\ \Omega$ in parallel with an inductance $L = 3.75$ H and two current sources

$$i_5 = 2\sin(5\omega t)\ \text{A}\ ;\quad i_7 = 2\sin(7\omega t)\ \text{A}$$

The load is supplied from a voltage source with four harmonics:

$$v_s = 14100\sin(\omega t) + 50\sin(3\omega t) + 150\sin(5\omega t) + 100\sin(7\omega t)\ ;\quad \omega = 376.99\ \text{rad/s}$$

via a feeder with the resistance $R_s = 3.0\ \Omega$ and inductance $L_s = 67$ mH. The load rms harmonic voltages and line currents listed in the Table 4.6 were determined using a simulation package.

The effect of a capacitor C, connected in parallel with the load, was observed by "measuring" the line rms current and the total harmonic distortion of the current, THD_I, in function of C, Fig. 4.14b. The graphs obtained demonstrate the ineffectiveness of the method based on the use of a single shunt capacitance. The uncompensated system ($C = 0$) has a line current $I_s = 10.96$ A. The minimum line current is 9.94 A at $C = 1.0\ \mu$F. At $C = 2.1\ \mu$F parallel resonance conditions are created. The optimum capacitance, estimated from (4.73), is $C_{opt} = 1.85\ \mu$F. This result is far from the best possible value. The effect of harmonic current amplification is reflected in the curve THD_I versus C; as C is increasing the current distortion is also increasing toward unacceptable levels. It can be shown that a single L–C tuned branch yields the best results at $m = 5.36$ with $C = 1.75\ \mu$F and $L = 0.14$ H. In this case the line current drops to $I_s = 7.22$ A and the current and voltage distortion become $THD_I = 22.6\ \%$ and $THD_V = 2.16\ \%$, respectively.

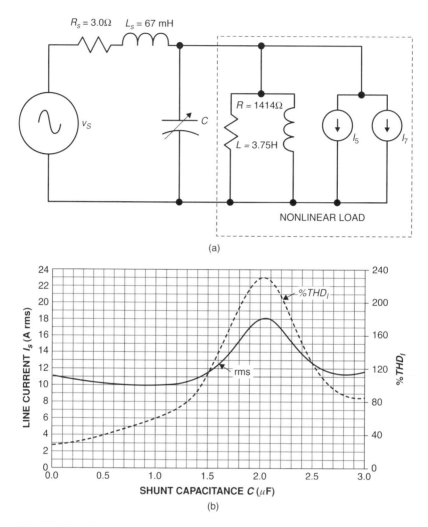

Figure 4.14 Nonlinear load compensated by a linear shunt capacitance: (a) Circuit. (b) RMS line current and THD_V vs. capacitance C.

An additional improvement is obtained if two tuned branches are used: For example if $C_5 = 1.0\ \mu F$ with $L_5 = 0.281$ H for $m = 5$ and $C_7 = 1.0\ \mu F$ with $L_7 = 0.143$ H for $m = 7$. Now the rms line current $I_s = 7.13$ A, $THD_I = 6.09$ % and $THD_V = 0.964$ %. These results are quite close to the best possible compensation.

The static compensators, while relatively inexpensive, do not offer the best practical solution. If the Thévenin's impedance of the system is changing, and if the background harmonic voltages are changing, such filters may become a liability and can be damaging or damaged by sinking harmonic currents injected by nonlinear loads supplied by the same feeder.

The purpose of this section is to discuss the basic concept of power factor compensation for nonsinusoidal situations. Modern power electronics enabled the development of dynamic compensators, known as active filters. The dominant school of thought uses the following approach: from Fryze's model we learn that the load current $i = i_a + i_b$, where i_a is the active,

Table 4.6   Simulation Results

h	1	3	5	7	THD (%)
V_h (V)	$9765\angle - 0.85°$	$344.8\angle - 5.3°$	$201.2\angle - 66.0°$	$407.6\angle - 88.8°$	5.84
I_h (A)	$9.76\angle - 45.87°$	$0.02\angle - 19°$	$1.43\angle - 5.6°$	$2.23\angle - 7.30°$	27.1

useful component and i_b is the parasitic, wattless one. If the compensator is made to operate like a current source that injects the current $i_{comp} = -i_b$, the line current $i_s = i + i_{comp} = i_a$ and the equivalent impedance of the load-compensator system becomes a simple resistance $R = 1/G = V^2/P$. The rms line current is minimized to $I_s = I_a$ and the current distortion equals the voltage distortion, $THD_I = THD_V$. This method can be called I_b-*compensation*.

Two phenomena happen when such a compensator is energized (the same thing happens with the well designed static compensators too): First, since the voltage drop across the supplying line decreases, the rms voltage at the load bus increases. Thus the utility may have to adjust the supply voltage. The second fact is that the load voltage spectrum is changing, usually is improved, and the load performance is affected for the better, unless the load is a plain heating element, or we deal with old-fashioned incandescent lamps.

Another power factor compensation strategy is to operate the compensator in such a manner that the line current is sinusoidal, $i_s = \widehat{I}_s \sin(\omega t + \gamma)$, the phase angle γ and the amplitude $\widehat{I}_s$ are adjusted to control the fundamental reactive and active powers $Q_1 = V_1 I_s \sin \gamma$ and $P = P_1 = V_1 I_s \cos \gamma$. This method can be called I_H-*compensation*.

This approach has the advantage not only that the power line loss is reduced, but that the electromagnetic interference caused by the supplying line may be significantly reduced.

Assuming a complete compensation with $I_H = 0$, $Q_1 = 0$ and $I_s = I_1$ results from (4.52) that

$$S^2 = \left(V_1^2 + V_H^2\right) I_s^2 = P^2 + D_V^2 = P^2 \left(1 + \frac{V_H^2 I_s^2}{V_1^2 I_s^2}\right) = P^2(1 + THD_V^2)$$

and

$$PF \approx \frac{1}{\sqrt{1 + THD_V^2}}$$

if $THD_V \leq 0.05$ results $PF \geq 0.9987$, which satisfies the most exigent expectations.

If $P_H < 0.02 P_1$, $TDH_V \leq 0.05$ and $THD_I \geq 0.40$, then (4.67) yields the approximation

$$PF \approx \frac{1}{\sqrt{1 + THD_I^2}} PF_1 \qquad (4.74)$$

The following numerical example will help us to better understand the power factor compensation theory for nonsinusoidal conditions. The studied system is sketched in Fig. 4.15a.

The voltage source

$$v_s = 330 \sin(\omega t) + 18 \sin(5\omega t) ; \quad \omega = 314.1 \text{ rad/s}$$

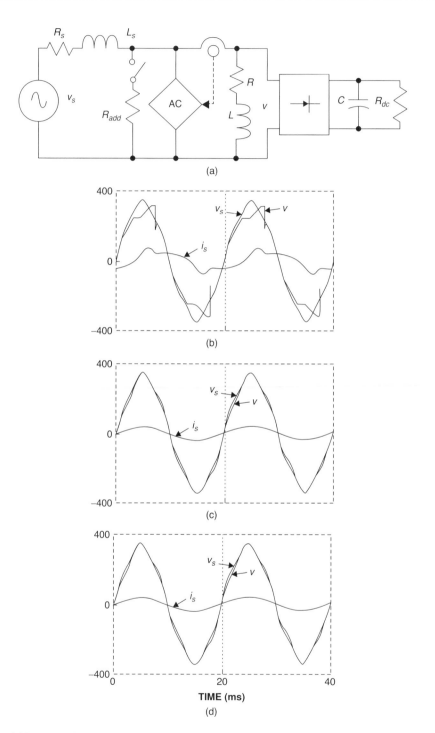

Figure 4.15   Example: (a) Circuit diagram. (b) Uncompensated circuit oscillograms. (c) Compensated circuit oscillograms for I_H-compensation. (d)I_b-compensation.

supplies a load that consists of a rectifier in parallel with a $R - L$ branch, ($L = 20$ mH and a resistance $R = 2.0$ Ω). The rectifier is a single-phase uncontrolled bridge supplying a filter capacitance $C = 1000$ μF and the dc load $R_{dc} = 25$ Ω. The load is connected to the source by a line with an equivalent resistance $R_s = 0.1$ Ω and inductance $L_s = 2.2$ mH. In Fig. 4.15b are shown the oscillograms of the voltage source, load voltage, and line current.

The results obtained from the simulation of this system are given in the Table 4.7:

Table 4.7 Simulation Results

	V (V)	I (A)	S (VA)	P (W)	PF	ΔP (W)	THD_V (%)	THD_I (%)
Uncompensated	210.15	42.84	9003.0	5068.8	0.563	183.53	13.93	30.74
I_b–Compensated	230.38	25.87	5960.0	5960.0	1.00	66.92	5.1	5.1
I_H–Compensated	231.75	26.83	6218.0	6189.7	0.995	71.98	5.49	0

The uncompensated load operates with 14% voltage distortion and the current distortion is near 31%.

Unity power factor is obtained when the load is I_b-compensated. In this case the compensated load (compensator and load) is equivalent to $G = P/V^2 = 5960/230.38^2 = 112.29$ mS, the rms line current decreases from 42.84 A to 25.87 A, and the line losses ΔP are drastically reduced. An impressive reduction of the current distortion is achieved. The rms voltage increases 9.6 %, from 210.15 V to 230.38 V. This may be a welcome change if the uncompensated system was operating at the lower limit of acceptable voltage, or it may be a problem that requires the utility intervention by readjusting the rms value of the voltage v_s.

The waveforms for I_H-compensation are shown in Fig. 4.15c. For the uncompensated system the dc load voltage is $V_{dc} = 275.94$ V and the power is $P_{dc} = 3045.7$ W. After compensation the dc output increases[7] to 294.38 V and 3466.38 W.

The 6.7% increase in the direct voltage and 13.8% increase in the output power may be inconsequential or may hinder the dc system operation if it was already operating at the upper limit of the admissible voltage.

The I_H-compensation yields very close results to I_b-compensation. The oscillograms in Figs. 4.15c and d are almost identical. For I_H compensation the line current becomes perfectly sinusoidal, $THD_I = 0$, and the load voltage distortion is only slightly higher than for I_b-compensation. In the general case without a thorough simulation and analysis of the local distribution system it is hard to decide what method will be more advantageous for the network when resonance, or near resonance, conditions exist. Resonances are caused by the interaction of shunt capacitors with line inductance or leakage inductance of transformers or voltage regulators. When implementing the I_b-compensation the equivalent conductance G tends to "compress" the resonance bell, hence attenuating the resonance effects. In situations where the line current spectrum must be maintained clear of harmonics the I_H-compensation may be preferred. Simulated oscillograms for I_b-compensation, Fig. 4.15d, are almost identical with the ones shown in Fig. 4.15c, for I_H-compensation. If the line rms current must be kept

---

[7] The dc load voltage is a function of the rectifier's topology, its type of filter and the dc load time constant. In this example the direct voltage is mainly affected by the peak voltage. This explains why the incremental changes of V_{dc} do not follow the changes of the rms voltage V.

constant an ideal resistor $R_{add} \approx 13 \; \Omega$, that simulates an additional unity power factor load, has to be connected at the main bus.

One learns from these observations that the concept of S implies a hypothetical ideal situation–not always possible to materialize–and in spite of the fact that S can be measured, S is only an indicator of what can be achieved under ideal conditions. Such an observation leads to an other important conclusion: The ideal conditions that define S should cover not only the optimum energy transfer to a single, isolated load, but should be extended to all the loads supplied by the same feeder.

4.8   Comments on Skin Effect, Apparent Power, and Power Factor

In all the above discussions the resistances were assumed constant and not affected by frequency. In practice the ac resistance is larger than the dc resistance, $R_{ac} > R_{dc}$, $R_{ac} = K_s R_{dc}$, where $K_s > 1$ is a coefficient accounting for the skin and proximity effects. In Fig. 4.16 are presented two curves that give the diameter of two solid cylindrical conductors with $K_s = 1.10$, as function of power source frequency. One conductor is made out of copper and the other of aluminum. The larger the diameter the lower is the frequency that will cause $K_s \geq 1.1$.

If we consider a resistance supplied with the nonsinusoidal voltage

$$v = \sum_h \widehat{V}_h \sin(h\omega t + \alpha_h) \tag{4.75}$$

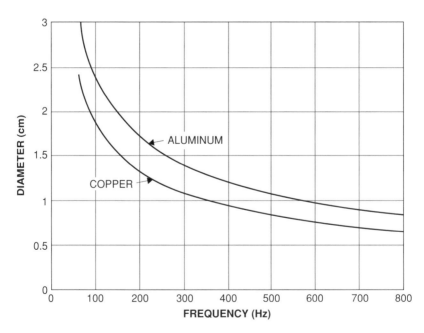

Figure 4.16   The diameter of a solid cylindrical conductor with skin effect coefficient $K_s = 1.10$ versus frequency.

the current will be

$$i = \sum_h \frac{\widehat{V_h}}{R_h} \sin(h\omega t + \alpha_h) \tag{4.76}$$

where $R_h = K_{sh} R_{dc}$, K_{sh} is the skin and proximity effect coefficient at the harmonic of order h. The apparent power squared can be derived from (4.3);

$$S^2 = \sum_{h=1}^{v} V_h^2 \sum_{h=1}^{v} \left(\frac{V_h}{R_h}\right)^2 = \left(\sum_{h=1}^{v} \frac{V_h^2}{R_h}\right)^2 + \sum_{m=1}^{v-1} \sum_{n=m+1}^{v} \left(\frac{V_m V_n}{R_n} - \frac{V_n V_m}{R_m}\right)^2 \tag{4.77}$$

Since $R_m \neq R_n$ the second term is not nil. The first term is the total active power

$$P = \sum_{h=1}^{v} \frac{V_h^2}{R_h} = P_1 + P_H \tag{4.78}$$

squared, and the second term is Budeanu's distortion power

$$D = \sqrt{\sum_{m=1}^{v-1} \sum_{n=m+1}^{v} \left(\frac{V_m V_n}{R_n} - \frac{V_n V_m}{R_m}\right)^2} \tag{4.79}$$

squared. One could also develop (4.77) according to (4.52)

$$S^2 = V_1^2 \left(\frac{V_1}{R_1}\right)^2 + V_1^2 \sum_{h \neq 1} \left(\frac{V_h}{R_h}\right)^2 + \left(\frac{V_1}{R_1}\right)^2 \sum_{h \neq 1} V_h^2 + \sum_{h \neq 1} \left(\frac{V_h^2}{R_h}\right)^2 + \sum_{\substack{m \neq n \\ m,n=1}} \left(\frac{V_m V_n}{R_n}\right)^2$$

$$= P_1^2 + D_I^2 + D_V^2 + P_H^2 + D_H^2 \tag{4.80}$$

Evidently in this case the reactive power is nil, $Q_1 = 0$, and the harmonic distortion power S_H, (4.59), lacks the terms $V_h I_h \sin \theta_h$, but all other types of nonactive powers are present.

Such resistances will rarely be encountered as particular stand alone loads, but can be frequently found in high current transformer windings, inductors and induction motor rotors, steel reinforced aluminum conductors and cables and conductors with large diameters.

The skin effect affects also inductances and mutual inductances value, but in a reversed way: as the frequency increases, the inductance decreases. The impact of frequency on inductance is less pronounced than the effect it has on resistance, (see problem 4.12).

An aspect that deserves special consideration stems from the fact that in practice there are many situations where conductors' resistance is strongly dependent on frequency. The skin effect should not be ignored when the penetration depth is less than the conductor radius, or half of the thickness of the conductor region that is exposed to perpendicular magnetic streamlines. Such systems deserve a thorough documentation based on the measurements carried by the manufacturer, or in-depth analysis meant to determine if the skin effect can be ignored or not.

Let us assume an I_H-compensated load supplied by a line with a heavier conductor, recognized to be affected by skin effect. The load is characterized by the apparent and active powers S and P and the power factor $PF = P/S$. The power loss in the supplying line is

$$\Delta P_{NS} = R_{sdc} \sum_h K_{sh} I_h^2 \tag{4.81}$$

Assuming next that the load is I_H-compensated and the line current I_s is sinusoidal with its rms value equal to the rms value of the nonsinusoidal current

$$I_s^2 = \sum_h I_h^2$$

The line power loss caused by I_s is in this case smaller than the power loss caused by the nonsinusoidal current with the same rms, thus

$$\Delta P_s = K_{s1} R_{sdc} I_s^2 = K_{s1} R_{sdc} \sum_h I_h^2 < \Delta P_{NS} \tag{4.82}$$

This result has an interesting interpretation: If the utility can provide a sinusoidal voltage and the consumers I_H-compensate the nonlinear loads, then the sinusoidal current I_s can have an rms value larger than the nonsinusoidal rms current value while the line power loss remains unchanged.

This conclusion leads to a skin effect corrected power factor. The equality of the power loss, $\Delta P_{NS} = \Delta P_s$, yields the value of a sinusoidal current that causes the same power loss as the nonsinusoidal current. From (4.81) and (4.82) it is obtained:

$$I_s = I_1 \sqrt{\sum_h \frac{K_{sh}}{K_{s1}} \left(\frac{I_h}{I_1}\right)^2} \tag{4.83}$$

The apparent power, i.e. the maximum active power that can be supplied to the receiving end of the line, is $S_s = V I_s > V I = S$, hence the power factor may be defined as

$$PF_s = \frac{P}{S_s} = \frac{S}{S_s} \frac{P}{S} = \frac{\sqrt{\sum_h \left(\frac{I_h}{I_1}\right)^2}}{\sqrt{\sum_h \frac{K_{sh}}{K_{s1}} \left(\frac{I_h}{I_1}\right)^2}} PF \tag{4.84}$$

A numerical example puts in perspective the above observation. We will consider a tubular copper conductor with a diameter $D = 5.64$ cm, a thickness $t = 0.2D = 1.13$ cm, carrying a current with the fundamental frequency of 60 Hz and the components:

$$I_1 = 2400 \text{ A}; \quad I_3 = 200 \text{ A}; \quad I_5 = 900 \text{ A} \quad \text{and} \quad I_7 = 350 \text{ A}$$

resulting in $TDH_1 = 41\%$.

The rms current is $I = 2594.7$ A with a current density $j = 2594.7/16 = 162$ A/cm$^2 \approx 1.0$ kA/in^2. Assuming a specific resistance of $1.72\ \mu\Omega$cm, the skin effect coefficients [27] are:

$$K_{s1} = 1.21 ; \quad K_{s3} = 2.09 ; \quad K_{s5} = 2.86 \quad \text{and} \quad K_{s7} = 3.18$$

From (4.84) results that the power factor has to be corrected with the coefficient

$$\frac{I}{\sqrt{\sum_h \frac{K_{sh}}{K_{s1}} I_h^2}} = \frac{2594.7}{2839.99} = 0.914$$

If we assume the same rms current with less distortion

$$I_1 = 2591.6\ \text{A} ; \quad I_3 = 20\ \text{A} ; \quad I_5 = 120\ A \quad \text{and} \quad I_7 = 35\ \text{A}$$

the current distortion drops to $TDH_I = 5.2\%$ and the power factor correction coefficient is an insignificant 0.998.

Of course if such a power factor study has to be done, one must weigh the skin effect affected power loss, in the involved conductors or windings, versus the entire feeder power loss. In situations where the conductor's resistance affects resonances and the degree of wave distortions, it is important to simulate correctly the inductors and resistors and not to ignore the skin and the proximity effects.

4.9 The Additiveness Problem

Let us assume N nonlinear loads connected at a point of common coupling and supplied with the nonsinusoidal voltage

$$v = \sum_{h \in v} \widehat{V}_h \sin(h\omega t + \alpha_h)$$

The instantaneous current supplied to a single load $k \in N$ is

$$i_k = \sum_{h \in v} i_{kh} = \sum_{h \in v} \widehat{I}_{kh} \sin(h\omega t + \alpha_{kh} - \theta_{kh}) ; \quad k \in N$$

Kirchhoff's current law applied to the point of common coupling leads to the total instantaneous current supplied to this cluster of loads:

$$i = \sum_{k=1}^{N} i_k = \sum_{k=1}^{N} \sum_{h \in v} i_{kh}$$

If we separate each harmonic current in the two basic components, the in-phase and the in-quadrature with the respective voltage harmonic, we have

$$i_{kh} = i_{kh}^C + i_{kh}^S$$

where[8]

$$i_{kh}^C = \widehat{I}_{kh} \cos(\theta_{kh}) \sin(h\omega t + \alpha_{kh})$$

and

$$i_{kh}^S = -\widehat{I}_{kh} \sin(\theta_{kh}) \cos(h\omega t + \alpha_{kh})$$

thus the total harmonic current of order h is:

$$i_h = \sum_{k=1}^N (i_{kh}^C + i_{kh}^S) = i_h^C + i_h^S$$

with

$$i_h^C = \widehat{I}_h \cos(\theta_h) \sin(h\omega t + \alpha_h) = \sum_{k=1}^N \widehat{I}_{kh} \cos(\theta_{kh}) \sin(h\omega t + \alpha_{kh})$$

and

$$i_h^S = -\widehat{I}_h \sin(\theta_h) \cos(h\omega t + \alpha_h) = \sum_{k=1}^N -\widehat{I}_{kh} \sin(\theta_{kh}) \cos(h\omega t + \alpha_{kh})$$

The total active rms harmonic current of order h supplied to all the N loads is

$$I_h \cos(\theta_h) = \sum_{k=1}^N I_{kh} \cos(\theta_{kh}) \tag{4.85}$$

and the reactive rms harmonic current of order h is

$$I_h \sin(\theta_h) = \sum_{k=1}^N I_{kh} \sin \theta_{kh} \tag{4.86}$$

---

[8] The superscripts C for $\cos(\theta_{kh})$ term and S for $\sin(\theta_{kh})$ term, are used to simplify the notations.

The total fundamental active power is

$$P_1 = \sum_{k=1}^{N} V_1 I_{k1} \cos(\theta_{k1}) = \sum_{k=1}^{N} P_{k1} \tag{4.87}$$

The total active harmonic power of order h is

$$P_h = \sum_{k=1}^{N} V_h I_{kh} \cos(\theta_{kh}) = \sum_{k=1}^{N} P_{kh} \tag{4.88}$$

and the total active harmonic power is

$$P_H = \sum_{h \in \nu} P_h \tag{4.89}$$

In the same way one finds the total fundamental reactive power

$$Q_1 = \sum_{k=1}^{N} V_1 I_{k1} \sin(\theta_{k1}) = \sum_{k=1}^{N} Q_{k1} \tag{4.90}$$

and the total reactive power of the h-order harmonic

$$Q_h = \sum_{k=1}^{N} V_h I_{kh} \sin(\theta_{kh}) = \sum_{k=1}^{N} Q_{kh} \tag{4.91}$$

All these friendly expressions prove the property of additiveness for the active and the reactive powers. They are the direct consequence of the power conservation law

$$vi = \sum_{k=1}^{N} vi_k$$

and for sinusoidal conditions they are most useful when the total apparent power required by N loads is computed:

$$S = \sqrt{\left(\sum_{k=1}^{N} P_k\right)^2 + \left(\sum_{k=1}^{N} Q_k\right)^2}$$

Under nonsinusoidal conditions this additiveness property applies to the active powers and the elementary reactive powers. (When Budeanu's method was discussed it was explained that it is wrong to use $Q_B = \sum_{k=1}^{N} Q_{kh}$.)

Unfortunately, when it comes to the remaining nonactive components, the magic of additiveness demonstrated in (4.87) to (4.91) is lost. As a matter of fact this was exactly W. Lyon's objection, (see section 4.1). However, Budeanu in a posthumously published work [28] proved that his distortion power, D_B can be separated in elementary distortion powers, each of them having the additiveness property and all of them being mutually orthogonal. He concluded that the total distortion power "obeys a property of vectorial conservation in a multidimensional space," this leading to the vectorial conservation of S.

This is a property of any nonactive power and can be easily demonstrated. We will focus on the current distortion power D_I (4.60):

$$D_I^2 = V_1^2 \sum_{\substack{h \in v \\ h \neq 1}} I_h^2 \tag{4.92}$$

and by separating the total rms harmonic current in its orthogonal terms

$$\begin{aligned}
I_h^2 &= [I_h \cos(\theta_h)]^2 + [I_h \sin(\theta_h)]^2 \\
&= \sum_{k=1}^{N} \left\{ [I_{kh} \cos(\theta_{kh})]^2 + [I_{kh} \sin(\theta_{kh})]^2 \right\} \\
&= \sum_{k=1}^{N} [(I_{kh}^C)^2 + (I_{kh}^S)^2]
\end{aligned} \tag{4.93}$$

we obtain for the total rms harmonic current squared:

$$\sum_{\substack{h \in v \\ h \neq 1}} I_h^2 = \sum_{k=1}^{N} \sum_{\substack{h \in v \\ h \neq 1}} [(I_{kh}^C)^2 + (I_{kh}^S)^2] \tag{4.94}$$

The current distortion squared of one load is:

$$D_{Ik}^2 = V_1^2 \sum_{\substack{h \in v \\ h \neq 1}} [(I_{kh}^C)^2 + (I_{kh}^S)^2] = \sum_{\substack{h \in v \\ h \neq 1}} [(D_{Ikh}^C)^2 + (D_{Ikh}^S)^2] \tag{4.95}$$

followed by the total current distortion squared:

$$D_I^2 = \sum_{k=1}^{N} \sum_{\substack{h \in v \\ h \neq 1}} [(D_{Ikh}^C)^2 + (D_{Ikh}^S)^2] = (D_I^C)^2 + (D_I^S)^2 \tag{4.96}$$

where

$$(D_I^C)2 = \sum_{k=1}^{N} \sum_{\substack{h \in v \\ h \neq 1}} (D_{Ikh}^C)^2 \quad \text{and} \quad (D_I^S)2 = \sum_{k=1}^{N} \sum_{\substack{h \in v \\ h \neq 1}} (D_{Ikh}^S)^2 \tag{4.97}$$

Thus, a vectorial additiveness holds true at the level of elementary powers. On a second thought we realize that D_V and D_I have $2(v-1)$ elementary components each. P_H and Q_H have $(v-1)$ components each and D_H has $2(v-1)(v-2)$. Counting two more components for P_1 and Q_1 we have a grand total of $2v^2$ components that do not make the concept of vectorial conservation very attractive.

The correct procedure for the computation of the total apparent power and its components should be based on the knowledge of the total current and bus voltage harmonic phasors at the point of common coupling. A numerical example will help explain this issue: Two loads connected in parallel are supplied with a nonsinusoidal voltage with the components:

$$V_1 = 100 \text{ V} ; \quad V_3 = 15 \text{ V} ; \quad V_5 = 20 \text{ V}$$

Load a with the harmonic current phasors $\mathbf{I}_h \angle \theta_h$:

$$\mathbf{I}_{a1} = 60\angle - 30° \text{ A} ; \quad \mathbf{I}_{a3} = 60\angle - 95° \text{ A} ; \quad \mathbf{I}_{a5} = 30\angle - 94° \text{ A}$$

and load b with

$$\mathbf{I}_{b1} = 80\angle - 45° \text{ A} ; \quad \mathbf{I}_{b3} = 30\angle 85° \text{ A} ; \quad \mathbf{I}_{a5} = 30\angle 88° \text{ A}$$

The following total powers are obtained: $P = P_a + P_b = 10.79$ kW, $P_H = P_{aH} + P_{bH} = -120.29 + 53.18 = -67.11$ W, $Q_1 = Q_{a1} + Q_{b1} = 8.657$ kvar, $D_I = 3.163$ kvar, $D_V = 3.457$ kvar, $D_H = 0.998$ kvar and $S = 14.68$ kVA. It is left to the reader, as a learning experience, to check the computations and to confirm that for all the discussed resolutions of S the additiveness of nonactive powers does not work; for example Fryze's method gives $Q_{aF} = 7.765$ kvar, $Q_{bF} = 7.015$ kvar while the total reactive power is $Q_F = 9.953$ kvar $\neq Q_{aF} + Q_{bF} = 14.780$ kvar.

4.10   Problems

4.1 A load consists of a capacitor with the susceptance $B_c = 105$ S connected in parallel with an inductance with susceptance $B_L = 225$ S. Both susceptances measured at ω rad/s. The load is supplied with a nonsinusoidal voltage

$$v = 141[\sin(\omega t) + 0.5\sin(5\omega t)] \text{ V}$$

Compute the instantaneous current and the instantaneous power. Compute S, P, Q_B, and D_B (according to Budeanu). You will find $Q_B = 0$ while there are present energy oscillations between the load and the source. Next, repeat the computations according to Czarnecki's model followed by Emanuel's.

4.2 A sinusoidal voltage $v = \widehat{V}\sin(\omega t)$ supplies a time-varying conductance $g = G_0[1 + a\cos(2\omega t + \alpha)]$, $0 < a < 1$. Determine Fryze's equivalent circuit and the expressions of i_a, i_b, P, Q_F, S, and PF. Repeat the procedure for different methods of S separation.

4.3 An inductance $L = 1$ H is supplied with $v = \widehat{V}[\sin(\omega t) + \sin(3\omega t)]$. Find the optimum value of the capacitance C that will minimize the line current. Find percent prospective reduction in line losses.

4.4 Return to problem 3.3 (Chapter 3). Compute S and its components according to Budeanu's, Czarnecki's, and Emanuel's methods. Find the PF.

4.5 Return to problem 3.9. Compute S and its components according to Budeanu's, Czarnecki's, and Emanuel's methods. Find the PF.

4.6 Return to problem 3.10. Compute S and its components according to Budeanu's, Czarnecki's, and Emanuel's methods. Find the PF.

4.7 Return to problem 3.11. Compute S and its components according to Budeanu's, Czarnecki's, and Emanuel's methods. Find the PF. Compare the results obtained for different methods.

4.8 A nonsinusoidal voltage V has the following rms harmonic current phasors $\mathbf{V}_1 = 40\angle 0°$ V, $\mathbf{V}_3 = 20\angle 80°$ V and $\mathbf{V}_5 = 10\angle 200°$ V. The voltage V supplies two loads, A and B, connected in parallel. The rms harmonic current phasors are: Load A : $\mathbf{I}_{A1} = 10\angle 45°$ A ; $\mathbf{I}_{A3} = 2\angle 50°$ A ; $\mathbf{I}_{A5} = 5\angle 215°$ A Load B : $\mathbf{I}_{B1} = 5\angle 45°$ A ; $\mathbf{I}_{B3} = 5\angle 50°$ A ; $\mathbf{I}_{B5} = 2\angle 215°$ A
Compute S_A, S_B and the total S, as well as their components according to different resolutions. Compare the results obtained for different methods.

4.9 A nonsinusoidal voltage $v = \sqrt{2}[\sin(\omega t) + \sin(3\omega t) + \sin(5\omega t)]$ supplies two branches in parallel; branch a has the impedance $1 + 0.5j$ at ω and branch b the impedance $1 - 0.25j$ at ω. Find i_a, i_b and $i = i_a + i_b$, the values and the components of S, S_a, and S_b.

4.10 Prove that the optimum capacitance of a series tuned filter $(m\omega L = 1/m\omega C)$, meant to minimize the rms current in a line supplying a nonlinear load with the voltage

$$v = \sum_h \widehat{V}_h \sin(h\omega t + \alpha_h)$$

is

$$C_{opt} = \frac{\displaystyle\sum_{h \neq m} \frac{hm^2 V_h I_h \sin\theta_h}{m^2 - h^2}}{\displaystyle\sum_{h \neq m} \left(\frac{hm^2}{m^2 - h^2} V_h\right)^2}$$

I_h is the h-order harmonic current in quadrature (lagging $90°$) with the h-order harmonic voltage.
Note: This result is correct only when the line impedance is negligible.

4.11 A resistance is supplied with a nonsinusoidal voltage

$$v = \sqrt{2}[240\sin(\omega t) + 30\sin(3\omega t) + 10\sin(5\omega t)] ; \quad \omega = 377 \text{ rad/s}$$

The dc resistance is $R_{dc} = 0.2 \, \Omega$ with the skin effect coefficients $K_{s1} = 1.09$, $K_{s3} = 2.3$ and $K_{s5} = 3.1$ Compute the apparent power, its components, and the power factor.

4.12 Repeat 4.11 assuming that the above resistance is distributed along the windings of an inductance $L_{dc} = 1.2$ mH.

At harmonic order h the inductance is $L_h = K_{Lh} L_{dc}$. The skin effect coefficients for the inductance are $K_{L1} = 0.97$, $K_{L3} = 0.92$ and $K_{L5} = 0.90$.

4.11   References

[1] Knowlton A. E.: "Reactive Powers in Need of Clarification," *Trans. AIEE,"* Vol. 52, Sept. 1933, pp. 744–805.

[2] Smith, L. P.: *"Mathematical Methods for Scientists and Engineers,"* Dover 1961, p. 309.

[3] Budeanu C. I.: *"Puissances Réactives et Fictives,"* Inst. National Roumain pour l'Étude de l'Aménagement et de l'Utilisation des Sources d'Énergie, Bucarest, 1927.

[4] Lagrange's Identity, Wikipedia, the free encyclopedia. http://en.wikipedia.org/wiki/Lagrange'sidentity.

[5] Yildirim D., Fuchs E. F.: "Commentary on Various Formulations of Distortion Power D," *IEEE Power Engineering Review,* May 1999, pp. 50–2.

[6] American Institute of Electrical Engineers, (ASA C42–1941), *"American Standard Definitions of Electrical Terms,"* August 12, 1941.

[7] *"The IEEE Standard Dictionary of Electrical and Electronics Terms,"* IEEE Std 100–1996, 6th edition, Institute of Electrical and Electronics Engineers, 1996.

[8] Arrillaga J., Bradley D. A., Bodger P. S.: *"Power System Harmonics,"* John Wiley, 1985, p. 123.

[9] Hart D. W.: *"Introduction to Power Electronics,"* Prentice Hall, 1997, p. 43.

[10] Acha E., Madrigal M.: *"Power System Harmonics,"* John Wiley, 2001, p. 39.

[11] Dugan R. C., McGranaghan M. F., Santoso S., Beaty H. W.: *"Electrical Power Systems Quality,"* McGraw-Hill, 2nd edition, 2003.

[12] Lyon W. V.: Discussion to the paper "Definitions of Power and Related Quantities," by H. L. Curtis and F. B. Silsbee, *Electrical Engineering*, Oct. 1935, p. 1121.

[13] Czarnecki L. S.: "What is Wrong with the Budeanu Concept of Reactive and Distortion Power and Why it Should be Abandoned," *IEEE Trans. on Instrumentation and Measurement*, Vol. 36, No. 3, Sept. 1987.

[14] Filipski P. S., Baghzouz Y., Cox M. D.: "Discussions of Power Definitions Contained in the IEEE Dictionary," *IEEE Transactions on Power Delivery,* Vol. 9, No. 3, July 1994, pp. 1237–44.

[15] IEEE Working Group on Nonsinusoidal Situations, "Practical Definitions for Powers in Systems with Nonsinusoidal Waveforms and Unbalanced Loads," *IEEE Transactions on Power Delivery,* Vol. 11, No. 1, Jan. 1996, pp. 79–101.

[16] GE Electrical Distribution and Control, *"Vector Electricity Meter with the Site Genie Monitor,"* Somersworth NH 03878, GEA 12601 2M 2/96, 1996.

[17] Fryze S.: "Effective, Wattless and Apparent Power in Circuits with Nonsinusoidal Waveforms of Current and Voltage," *Elektrotechnishe Zeitschrift,"* No. 25, June 23, 1932, pp. 596–99, 625–27, 700–702.

[18] Depenbrock M.,*"Investigation of Voltage and Power Conditions at Converters without Energy Storage,"* Ph.D. Thesis, T. H. Hanover, 1962.

[19] Depenbrock M.: *"Active and Fictitious Power of Periodic Currents in Single-Phase and Polyphase Systems with Periodical Voltages of any Waveforms,"* ETG Fachberichteg, Berlin und Offenbach, VDE – Verlag 1979, pp. 17–59. 1962.

[20] Czarnecki L. S.: "Considerations on the Reactive Power in Nonsinusoidal Situations," *IEEE Trans. on Instrumentation and Measurement*, Vol. 34, No. 3, Sept. 1984, pp. 399–404.

[21] Emanuel A. E.: "On the Assessment of Harmonic Pollution," *IEEE Trans. on Power Delivery*, Vol. 10, No. 3, July 1995, pp. 1693–98.

[22] IEEE Power Engineering Society, Power System Instrumentation and Measurements Committee:*"IEEE Standard Definitions for the Measurement of Electric Power Quantities Under Sinusoidal, Nonsinusoidal, Balanced, or Unbalanced Conditions,"* IEEE, February 2010.

[23] Emanuel A. E.: "Reflections on the Effective Voltage Concept," *Proceedings of the 6th International. Workshop on Power Definitions and Measurements*, Milan, Italy, Oct. 2003, pp. 1–7.

[24] Erlicki M. S., Emanuel A. E.: "New Aspects of Power Factor Improvement," *IEEE Trans. on Industry applications*, Vol. 4, No. 4, July/Aug. 1968, pp. 441–55.

[25] Shephard W., Zand P.: *"Energy Flow and Power Factors in Nonsinusoidal Circuits,"* Cambridge University Press, Cambridge, 1979, p. 191.

[26] Kuster N. L., Moore W. J. M.: "On the Definition of Reactive Power under Nonsinusoidal Conditions," *IEEE Trans. on Power Apparatus and Systems*, Vol. 99, Sept./Oct. 1980, pp. 1845–54.

[27] *"IEEE Standard for Metal-Enclosed Bus,"* ANSI/IEEE C37. 23 – 1987, p. 23.

[28] Budeanu C. I.: "The Physical Role of Instantaneous Quantities in Conservation Phenomena," *Energetica*, Vol. 34, No. 6, 1986, pp. 277–82. pp. 1845–54.

[29] Depenbrock M.: "The FBD-Method, A Generally Applicable Tool for Analyzing Power Relations," *IEEE Trans. on Power Systems*, Vol. 8, No. 2, May 1993, pp. 381–87.

5

Three-Phase Systems with Sinusoidal Waveforms

Whoever undertakes to set himself up as a judge in the field of Truth
and Knowledge is ship wrecked by the laughter of gods.

—Albert Einstein *Aphorisms for Leo Baeck*

Three-phase alternators provide nearly perfect symmetrical sinusoidal voltages. Transmission and distribution lines have almost identical impedances on all three phases. Three-phase equipment, motors, rectifiers, filters, and transformers are balanced systems designed to operate with 120° shifted, but identical voltages, on all three phases. Smaller loads like lamps, office equipment, and small motors are single-phase loads; nevertheless, when large clusters of single-phase loads are supplied by three-phase circuits, the engineers group the loads in such a manner as to ensure minimum neutral current. Balanced loads help to reduce line power loss, minimizing the negative- and the zero-sequence voltages, thus leading to better motor efficiency, longer life span, and minimization of atypical harmonics' injection. Of course there are large single-phase loads that, when integrated in the three-phase system, cause significant imbalance: trains driven by single-phase motors or welding machines are typical examples. Only in computer simulations and well equipped laboratories does one encounter absolutely perfect balanced three-phase systems. In practice traces of negative- and zero-sequence voltages and relatively larger amounts of negative- and zero-sequence currents are always present.

While the definition of apparent power and its components in balanced three-phase systems with sinusoidal waveforms is uncontested, the definitions for unbalanced conditions are still a source of heated debate. At the 36th Annual Convention of AIEE in 1920 a special Joint Committee [1,2,3] reported the following:

"The subject of power factor in polyphase circuits has been the center of increasing discussions in recent years. No agreement has yet been reached upon a definition ... Until recent years, most polyphase loads were approximately balanced, while the differences between various possible definitions of power factor become of importance only in unbalanced loads...

Power Definitions and the Physical Mechanism of Power Flow   Alexander Eigeles Emanuel
© 2010 John Wiley & Sons, Ltd

The increasing commercial importance of this character of load and the growing tendency toward such refinements in power contracts and rates as will reflect accurately the various elements entering into the cost of service, have combined to render this power factor problem a matter of immediate and urgent practical importance."

The polyphase systems are today more balanced and predictable than 90 years ago; at the same time power engineers expect higher efficiencies, better power factors, and reliability of service. Numerous conventions followed the 1920 event, and every decade was marked by some progress on understanding the physics of energy flow and on the capability to build more accurate and versatile metering instrumentation. One thing did not change: the fact that the debate for a universally accepted definition of apparent power and power factor for the unbalanced systems still continues today. The observation highlighted in the AIEE report is as true today as it was in 1920.

This chapter deals with three-phase balanced and unbalanced linear systems with sinusoidal voltage and current waveforms. The flow of energy under sinusoidal conditions will be analyzed and different apparent power definitions compared.

5.1   Background: The Balanced and Symmetrical System

We start assuming a balanced three-phase load supplied with the instantaneous line-to-neutral voltages and line currents

$$
\begin{aligned}
v_a &= \widehat{V}\sin(\omega t + \alpha)   &   i_a &= \widehat{I}\sin(\omega t + \alpha - \theta) \\
v_b &= \widehat{V}\sin(\omega t - 120° + \alpha)   &   i_b &= \widehat{I}\sin(\omega t - 120^0 + \alpha - \theta) \\
v_c &= \widehat{V}\sin(\omega t + 120° + \alpha)   &   i_c &= \widehat{I}\sin(\omega t + 120^0 + \alpha - \theta)
\end{aligned}
\tag{5.1}
$$

The instantaneous power

$$
p = v_a i_a + v_b i_b + v_c i_c = p_p + p_q
$$

has an active component

$$
\begin{aligned}
p_p &= VI\cos(\theta)[1 - \cos(2\omega t + 2\alpha)] + VI\cos(\theta)[1 - \cos(2\omega t - 240° + 2\alpha)] \\
&+VI\cos(\theta)[1 - \cos(2\omega t + 240^0 + 2\alpha)] = 3VI\cos(\theta) = P
\end{aligned}
\tag{5.2}
$$

and a reactive component

$$
\begin{aligned}
p_q &= -VI\sin(\theta)\sin(2\omega t + 2\alpha) - VI\sin(\theta)\sin(2\omega t - 240° + 2\alpha) \\
&-VI\sin(\theta)\sin(2\omega t + 240^0 + 2\alpha) = 0
\end{aligned}
\tag{5.3}
$$

We obtained two remarkable results: First, equation (5.2) indicates that the intrinsic powers cancel each other, i.e. a three-phase alternator under steady-state conditions delivers a perfect constant power and the prime mover driving the alternator delivers a constant torque void of oscillations. Similarly, an ideal induction or synchronous motor supplied with symmetrical voltages produces a constant torque. The energy supplied by a symmetrical three-phase voltage to a balanced system is $W = Pt$. No oscillating term is to be found in this expression. However, this is only the "macroscopic" view, just a mathematical conclusion resulting from (5.2).

The second result is even more puzzling; according to (5.3) there is a perfect cancellation among the reactive instantaneous powers. Does this means there is no reactive power Q? Does this means that there are no oscillations between the inductive or capacitive loads and the three-phase sources, as claimed by some researchers [4]?

From (5.2) and (5.3) it is also clear that each phase contributes with its own active power

$$P_a = P_b = P_c = VI\cos(\theta)$$

reactive power

$$Q_a = Q_b = Q_c = VI\sin(\theta)$$

and apparent power

$$\mathbf{S}_a = \mathbf{S}_b = \mathbf{S}_c = \mathbf{V}\,\mathbf{I}^* = P_a + jQ_a \quad \text{with} \quad |S_a| = |S_b| = |S_c| = VI$$

where the current phasor $\mathbf{I}^* = I\angle\theta$ is the conjugate of the line current phasor $\mathbf{I} = I\angle -\theta$.

Assuming identical lines' resistances R_s, the total power loss in the three supplying lines is:

$$\Delta P = \Delta P_a + \Delta P_b + \Delta P_c = 3R_sI^2 = \frac{3R_s}{V^2}(P_a^2 + Q_a^2) = \frac{3R_s}{V^2}S_a^2 \tag{5.4}$$

Equation (5.4) proves the existence of a total reactive power $Q = 3Q_a$ manifested through the power loss $3R_sQ^2/V^2$.

The total apparent power of the balanced load is

$$S = 3VI = \sqrt{3}V_{LL}I = \sqrt{P^2 + Q^2}\,; \quad V_{LL} = \sqrt{3}\,V$$

or in a complex form

$$\mathbf{S} = 3\mathbf{V}\,\mathbf{I}^* = 3V\angle\alpha\, I\angle(\theta - \alpha) = 3VI[\cos(\theta) + j\sin(\theta)] = P + jQ \tag{5.5}$$

where

$$P = 3P_a = \Re\{\mathbf{VI}^*\} = S\cos(\theta) = 3VI\cos(\theta) \tag{5.6}$$

is the total active power and

$$Q = 3Q_a = \Im\{\mathbf{VI}^*\} = S\sin(\theta) = 3VI\sin(\theta) \tag{5.7}$$

is the total reactive power.

The oscillations of instantaneous active and reactive powers of the phases a, b, and c are not illusions, as claimed in [4], they do exist. Equation (5.3) presents the overall result, it hides the physical details of the electromagnetic field, it does not model the distribution of power density in the space that surrounds the conductors. If the reactive powers Q_a, Q_b, and Q_c cause Joule and eddy-current losses in the conductors, then a reciprocating electromagnetic wave must run along each conductor between the source and the L or C of the load. A small component of these electromagnetic waves impinges into conductors' surface and carry the energy dissipated by the conductors in heat. The existence and nature of power oscillations in three-phase systems is discussed and proved with the help of Poynting vector theory in Appendix IV.

5.2   The Three-Phase Unbalanced System

When the power system is not balanced the general expressions of the instantaneous line currents and line-to-line voltages are as follows:

$$i_a = \widehat{I_a}\ \sin(\omega t + \alpha_a - \theta_a) \tag{5.8}$$
$$i_b = \widehat{I_b}\ \sin(\omega t - 120^0 + \alpha_b - \theta_b) \tag{5.9}$$
$$i_c = \widehat{I_c}\ \sin(\omega t + 120^0 + \alpha_c - \theta_c)\ ;\ \ \theta_a \neq \theta_b \neq \theta_c\ ;\ \ \alpha_a \neq \alpha_b \neq \alpha_c \tag{5.10}$$
$$v_a = \widehat{V_a}\ \sin(\omega t + \alpha_a) \tag{5.11}$$
$$v_b = \widehat{V_b}\ \sin(\omega t - 120^\circ + \alpha_b) \tag{5.12}$$
$$v_c = \widehat{V_c}\ \sin(\omega t + 120^\circ + \alpha_c) \tag{5.13}$$

The total instantaneous power is

$$p = v_a i_a + v_b i_b + v_c i_c = p_a + p_b + p_c$$

$$p_a = V_a I_a \cos\theta_a [1 - \cos(2\omega t + 2\alpha_a)] - V_a I_a \sin\theta_a \sin(2\omega t + 2\alpha_a)$$
$$p_b = V_b I_b \cos\theta_b [1 - \cos(2\omega t + 2\alpha_b - 240^\circ)] - V_b I_b \sin\theta_b \sin(2\omega t + 2\alpha_b - 240^\circ)$$
$$p_c = V_c I_b \cos\theta_c [1 - \cos(2\omega t + 2\alpha_c + 240^\circ)] - V_c I_c \sin\theta_c \sin(2\omega t + 2\alpha_c + 240^\circ)$$

Each phase carries an active power

$$P_a = V_a I_a \cos(\theta_a) \quad P_b = V_b I_b \cos(\theta_b) \quad P_c = V_c I_c \cos(\theta_c)$$

We also observe the existence of intrinsic power appertaining to each of the active powers. This time the instantaneous intrinsic powers do not completely cancel each other, nevertheless the residual intrinsic power does not cause any additional power loss in the line.

The reactive components of the current cause oscillations with the amplitudes Q_a, Q_b, and Q_c that define the per phase reactive powers

$$Q_a = V_a I_a \sin(\theta_a) \quad Q_b = V_b I_b \sin(\theta_b) \quad Q_c = V_c I_c \sin(\theta_c)$$

For each one of the three phases, a, b, and c, we can define a power triangle $S_a\ P_a\ Q_a$, $S_b\ P_b\ Q_b$, and $S_c\ P_c\ Q_c$, respectively, where $S_a = V_a I_a$, $S_b = V_b I_b$ and $S_c = V_c I_c$, Fig. 5.1. This geometrical representation leads to two apparent power definitions:

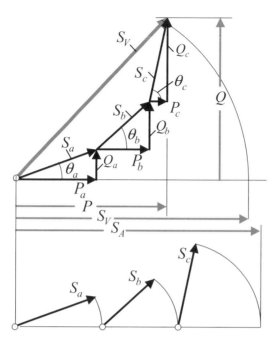

Figure 5.1 Vector apparent power S_V and arithmetic apparent power S_A: Geometrical interpretation by means of power triangles, (sinusoidal conditions).

1. The *Arithmetic Apparent Power*:

$$S_A = S_a + S_b + S_c \tag{5.14}$$

yielding an arithmetic power factor

$$PF_A = \frac{P}{S_A} \quad \text{where} \quad P = P_a + P_b + P_c$$

2. The *Vector Apparent Power*:

$$\mathbf{S_V} = \mathbf{V}_a\mathbf{I}_a^* + \mathbf{V}_b\mathbf{I}_b^* + \mathbf{V}_c\mathbf{I}_c^* = P_a + P_b + P_c + j(Q_a + Q_b + Q_c) = P + jQ \tag{5.15}$$

$$S_V = \sqrt{P^2 + Q^2}$$

leading to the vector power factor

$$PF_V = \frac{P}{S_V}$$

When the monitored three-phase system is balanced both definitions give identical results. However, for unbalanced conditions the results differ: $S_A \geq S_V$, $PF_A \leq PF_V$ and the truth is that both definitions, S_A and S_V, do not fulfill the cardinal requirement of *"line power loss proportionality to S^2"* (see (2.32)), and both definitions give incorrect apparent power and power factor values.

A better definition was proposed by F. Buchholz [5] and explained in 1933 by W. M. Goodhue [6,7]. Buchholz introduced the notion of *system rms or effective voltage*:

$$\sqrt{3}V_e = \sqrt{\frac{V_{ab}^2 + V_{bc}^2 + V_{ca}^2}{3}} \tag{5.16}$$

and *system rms or effective current*:

$$I_e = \sqrt{\frac{I_a^2 + I_b^2 + I_c^2}{3}} \tag{5.17}$$

Leading to the *effective apparent power*

$$S_e = 3V_e I_e \tag{5.18}$$

Goodhue pointed to the fact that (5.16) and (5.17) are easily derived from the condition of equivalence of the actual power loss with the hypothetical equivalent system power loss. The current dependent losses are

$$\Delta P_I = R_s(I_a^2 + I_b^2 + I_c^2) = 3R_s I_e^2 \tag{5.19}$$

A more complete definition discussed in [8] takes into account the fact that the neutral current path also contributes to the total power loss and the possibility that the four conductors may have different resistances. In this case the equivalence is

$$\begin{aligned} \Delta P_I &= R_a I_a^2 + R_b I_b^2 + R_c I_c^2 + R_n I_n^2 \\ &= R_e(I_a^2 + I_b^2 + I_c^2 + I_n^2) = 3R_e I_e^2 \end{aligned} \tag{5.20}$$

and

$$I_e = \sqrt{\frac{I_a^2 + I_b^2 + I_c^2 + I_n^2}{3}} \tag{5.21}$$

The equivalent resistance R_e is a function of the actual resistances R_a, R_b, R_c, and R_n and the four currents (see Appendix VII). From (5.20) results:

$$R_e = \frac{R_a I_a^2 + R_b I_b^2 + R_c I_c^2 + R_n I_n^2}{I_a^2 + I_b^2 + I_c^2 + I_n^2}$$

A common case is when $R_a = R_b = R_c = R_s$ and $R_n = \rho R_s$, then (5.21) becomes

$$I_e = \sqrt{\frac{I_a^2 + I_b^2 + I_c^2 + \rho I_n^2}{3}} \tag{5.22}$$

The equivalent or effective voltage can be found by considering the voltage dependent losses (core-losses and dielectric) and assuming an equivalent shunt resistance R_s. These voltage dependent losses are

$$\Delta P_V = \frac{V_{ab}^2 + V_{bc}^2 + V_{ca}^2}{R_s} = \frac{9 V_e^2}{R_s} \tag{5.23}$$

yielding

$$V_e = \sqrt{\frac{V_{ab}^2 + V_{bc}^2 + V_{ca}^2}{9}} \tag{5.24}$$

There are load topologies where (5.24) gives small errors. A detailed discussion of (5.24) is found in section 5.5.4. The IEEE Std. 1459–2010 approach to V_e takes into consideration the conservation of the active power supplied to the load instead of the conservation of voltage dependent losses (see section 5.5.4).

5.3 The Power Factor Dilemma

Polyphase systems are designed in mind as one entity, one "pipeline" meant to transmit electric energy from the generators to substations and consumers. The energy supplier expects and plans for energy transfer conditions that are conducive to reduced invested capital and operating cost, conditions that are not causing equipment damage or early aging, conditions that ensure proper voltage regulation and minimum interference with other utilities that share the same right of way (communication, gas, and water). This means two major requirements:

- Maximum utilization of conductor cross-sectional area without causing insulation or mechanical damage.
- Minimum neutral (residual) current that may stray through low impedance paths and cause raised potential of improperly grounded items and equipment as well as instrumentation disturbances due to common mode electromagnetic interference.

The simplified concept of a three-phase line utilization is depicted in Fig. 5.2. The four conductors are represented by the four circles. The cross-sectional area is shown proportional to the $I_{a,b,c,n}^2$. The cross-sections are divided according to the active and reactive current squared. One may replace the sectors dedicated to active and reactive currents with P^2 and Q^2 respectively. In an unbalanced system (Fig. 5.2a), the apparent and active power distribution among the conductors is not even. The balanced systems with $PF < 1.0$ have perfect symmetry, Fig. 5.2b, but the conductor's utilization is not complete. Only when $PF = 1.0$ is the entire

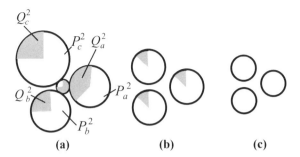

Figure 5.2   The utilization of a three-phase, four-wire feeder: (a) Unbalanced conditions. Total conductors' cross-sectional area is proportional to the total apparent power squared. The darker sectors are proportional to the nonactive power squared. (b) Balanced system with $PF < 1.0$. (c) Balanced System with $PF = 1.0$.

cross-sectional area of each phase is fully utilized by carrying only active power. Each phase carries the same amount of power, Fig. 5.2c.

The neutral wire is an auxiliary conductor, its main role is to reduce transient voltages and to carry three times zero-sequence current during unsymmetrical faults. In three-phase four-wire unbalanced systems the neutral conductor carries the zero-sequence current, but a negligible amount of power. Continuous current flow through the neutral conductor opens the door to a multitude of consumer and energy supplier annoyances. An ideal polyphase system with unity power factor must operate with zero neutral current [9].

The first to address effectively the definition of power factor in three-phase systems was W. V. Lyon [1]. In 1920 he had extended the single-phase approach to polyphase systems; his recommendation was *"to imagine that the actual load is replaced by an arrangement of 'standard' circuits, i.e., constant non reactive resistances - so adjusted that the root-mean-square line currents they take are equal to these taken by the actual load. [and] The rms line voltages are the same in each case [for the original and for the 'imagined circuit]."*

In the 1920s and 1930s the pillars of the electrical engineering community, charmed by the elegance of Fortescue's symmetrical components theory, which fitted hand-in-glove with the vector apparent power definition, dismissed Lyon's recommendation. Symmetrical components help to gain insight into the structure of the effective current and voltage and play a significant role when different apparent power definitions are evaluated, one against the other. However, symmetrical components alone, without a correct interpretation of the physical mechanisms of energy transmission and conversion, cannot lead to a correct apparent power definition. (Section 5.4 is completely dedicated to symmetrical components.) Meanwhile a few numerical examples will shed light on the power factor problem.

Example 5.1   A three-phase bus a, b, c, n is supplied by a four-wire system. The line-to-neutral bus voltages are $V_a = \mathbf{V}\angle 0°$, $V_b = \mathbf{V}\angle -120°$, and $V_c = \mathbf{V}\angle 120°$. A resistance R is connected between the terminal a and the neutral n.

We deal with an unbalanced load where only phase a supplies the active power

$$P = P_a = S_a = \frac{V^2}{R} = VI \; ; \quad I = \frac{V}{R}$$

and the other two phases remain disconnected

$$P_b = P_c = 0 \quad S_b = S_c = 0$$

In this case we obtain the total apparent powers $S_A = S_V = S_a = P$ and the power factors $PF_A = PF_V = P/S = 1.0$. Nevertheless, the power factor cannot be unity if only 33% of the active material is used.

Now, let us consider the apparent power definition (5.18), given by Buchholz and Goodhue; the effective voltage and current are

$$V_e = V \quad \text{and} \quad I_e = \sqrt{\frac{I^2 + I^2}{3}} = \sqrt{\frac{2}{3}}I$$

hence

$$S_e = 3\sqrt{\frac{2}{3}}VI \quad \text{and} \quad PF_e = \frac{P}{S_e} = \frac{1}{\sqrt{6}} = 0.408$$

The power factor indicates how well the conductors are utilized. In single-phase systems P/S means that $(P/S)^2$ of the conductor cross-section is used to carry active power. This is tantamount with the definition $PF = \sqrt{\Delta P_c / \Delta P}$, where ΔP are the actual losses and ΔP_c are the losses with the load compensated to unity power factor. Assuming the same resistance R_s for each one of the lines as well as for the neutral path, results

$$\Delta P = 2R_s I^2$$

For the compensated feeder the currents are symmetrical, $I'_a = I'_b = I'_c = I/3$, delivering the same power $P = VI' = 3VI/3$. From here one finds the power loss in the compensated feeder:

$$\Delta P_c = 3R_s(I/3)^2 = R_s I^2/3$$

and the power factor $PF = \sqrt{\Delta P_c / \Delta P} = \sqrt{1/6}$.

Example 5.2 We connect now the resistance R line-to-line between the terminals a and b. This unbalanced load takes the active power

$$P = \frac{3V^2}{R} = \sqrt{3}VI; \quad I = \frac{\sqrt{3}V}{R}$$

with the line current phasors (Fig. 5.3a) $\mathbf{I}_a = I\angle 30°$ and $\mathbf{I}_b = I\angle - 150°$. Thus

$$P_a = P_b = VI\cos(30°)$$
$$Q_a = VI\sin(-30°) \quad \text{and} \quad Q_b = VI\sin(30°)$$
$$S_a = VI\angle - 30° = VI\cos(30°) - _J VI\sin(30°)$$
$$S_b = (V\angle - 120°)(I\angle 150°) = VI\cos(30°) + _J VI\sin(30°)$$

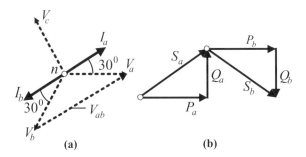

Figure 5.3   Example 2: (a) Phasor diagram. (b) Power triangles.

We have $P_a = P_b = 0.866VI$, $P_c = 0$, $Q_b = -Q_a = 0.5VI$, $Q_c = 0$, $S_a = S_b = VI$ and $S_c = 0$. The power triangles, Fig. 5.3b, yield

$$S_V = P_a + P_b = P = \sqrt{3}VI \quad \text{and} \quad PF_V = \frac{P}{S_V} = 1.0$$

$$S_A = S_a + S_b = 2VI \quad \text{and} \quad PF_A = \frac{P}{S_A} = \frac{\sqrt{3}VI}{2VI} = 0.866$$

Since the load is purely resistive and only two-thirds of the active material is used and neither one of the computed power factors is correct.

The Buchholz-Goodhue method based on $V_e = V$ and $I_e = \sqrt{2/3}I$ yields

$$S_e = \sqrt{6}\,VI \quad \text{and} \quad PF_e = \frac{P}{S_e} = \frac{\sqrt{3}\,VI}{\sqrt{6}\,VI} = \frac{1}{\sqrt{2}} = 0.707$$

Again if we go to the very meaningful definition $PF = \sqrt{\Delta P_c / \Delta P}$, and we use the compensated feeder line current

$$I' = P/3V = \sqrt{3}VI/3V = I/\sqrt{3}$$

to find the compensated line losses, $\Delta P_c = 3R_s I^2/3$. The actual losses are $\Delta P = 2R_s I^2$ yielding the same power factor expression, $PF = \sqrt{R_s I^2/(2R_s I^2)} = 0.707$.

Example 5.3   We will assume a more involving load: The resistance R connected a to n, an inductance with the reactance jR connected b to n and a capacitance with the reactance $-jR$ connected c to n, Fig. 5.4a. The line current phasors are

$$\mathbf{I}_a = \frac{V}{R} = I \quad \mathbf{I}_b = \frac{V\angle -120°}{jR} = I\angle -210° \quad \mathbf{I}_c = \frac{V\angle 120°}{-jR} = I\angle 210°$$

The phasor diagram is shown in Fig. 5.4b. The power triangles are reduced to three segments, Fig. 5.4c, $P_a = S_a = VI$, $Q_a = 0$, $P_b = P_c = 0$, $Q_b = S_b = VI$ and $Q_c = S_c = VI$. From here results $P = P_a = VI$ and for the vector apparent power approach:

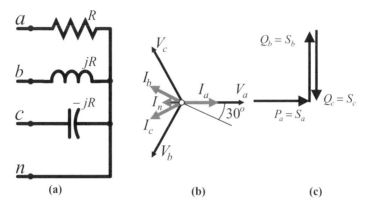

Figure 5.4 Example 3: (a) Connections. (b) Phasor diagram. (c) Power vectors.

$$S_V = S_a = VI \qquad PF_V = \frac{P}{S_V} = 1.0$$

For the arithmetic apparent power approach:

$$S_A = S_a + S_b + S_c = 3VI \qquad PF_A = \frac{P}{S_A} = \frac{1}{3}$$

In this case only one third of the active material is utilized, two conductors are "misused," and a unity power factor is not possible.

If the neutral current is ignored, or if $\rho \to 0$, then $I_e = \sqrt{(I^2 + I^2 + I^2)/3} = I$ leading to

$$S_e = 3VI \quad \text{and} \quad PF_e = \frac{P}{S_e} = \frac{VI}{3VI} = 0.333$$

If the neutral current $\mathbf{I}_n = (\sqrt{3} - 1)I \angle 180°$ is not ignored and $\rho = 1$, then $I_e = \sqrt{1 + 0.178I} = 1.086\,I$ and $S_e = 3.257\,VI$ yielding $PF_e = 0.307$.

It is left to the reader to check this result using the $PF = \sqrt{\Delta P_c/\Delta P}$ approach.

5.4   Powers and Symmetrical Components

5.4.1   How Symmetrical Components are Generated

For all practical purposes the voltages and currents at the generator's bus, under normal operation, are clean positive-sequence voltages and currents. It is appropriate to ask the following question: how the negative- and zero-sequence components are generated if all the energy sources belonging to this system do not generate this type of energy? The answer is: unbalanced loads convert positive-sequence energy into negative- and zero-sequence energy, just like a nonlinear load converts some input energy at the power system frequency in energy

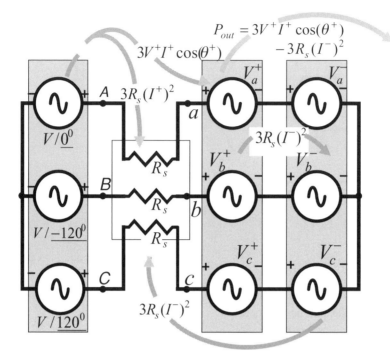

Figure 5.5   Active power flow: Three-phase, three-wire system with unbalanced load.

at higher harmonics frequencies. A simple example will help to explain the basic energy flow mechanism particular to such a situation. In Fig. 5.5 a three-phase symmetrical voltage $V\angle 0°$, $V\angle -120°$ and $V\angle 120°$ supplies an unbalanced load via a three-wire line with three equal resistances R_s. The load is equivalent to three positive-sequence voltages each in series with a negative-sequence voltage.

The source produces the following instantaneous power

$$p_{Gen} = v_A i_a^+ + v_B i_b^+ + v_C i_c^+ = R_s[(ia^+)^2 + (ib^+)^2 + (ic^+)^2] + p_{\substack{Suppl \\ Load}} \qquad (5.25)$$

where $v_{A,B,C}$ are the line-to-neutral instantaneous voltage of the source, $i_{a,b,c}^+$ are the load's positive-sequence instantaneous current, and $p_{\substack{Suppl \\ Load}}$ is positive-sequence power supplied to the load by the generator. The instantaneous power measured at the load terminals a, b, and c is

$$p_{out} = v_a^+ i_a^+ + v_b^+ i_b^+ + v_c^+ i_c^+ + v_a^- i_a^- + v_b^- i_b^- + v_c^- i_c^- \qquad (5.26)$$

where $v_{a,b,c}^+$ are the load's positive-sequence instantaneous voltage and $v_{a,b,c}^-$, are the load's negative-sequence instantaneous voltage.

Kirchhoff's voltage law written for any of the three phases positive- and negative-sequence component is

$$v_K = v_k^+ + R_s i_k^+ \quad \text{and} \quad v_k^- + R_s i_k^- = 0; \quad K = A, B, C; \quad k = a, b, c$$

Substitution of these two equations in (5.26) gives

$$
\begin{aligned}
p_{out} &= (v_A - R_s i_a^+) i_a^+ + (v_B - R_s i_b^+) i_b^+ + (v_C - R_s i_c^+) i_c^+ \\
&\quad + (-R_s)[(i_a^-) i_a^- + (i_b^-) i_b^- + (i_c^-) i_c^-] \\
&= v_A i_a^+ + v_B i_b^+ + v_C i_c^+ \\
&\quad - R_s[(i_a^+)^2 + (i_b^+)^2 + (i_c^+)^2] - R_s[(i_a^-)^2 + (i_b^-)^2 + (i_c^-)^2]
\end{aligned} \tag{5.27}
$$

Comparing (5.27) with (5.25) we note that $p_{\substack{Suppl \\ Load}} > p_{out}$ and the difference

$$p_{\substack{Suppl \\ Load}} - p_{out} = R_s[(i_a^-)^2 + (i_b^-)^2 + (i_c^-)^2] = -(v_a^- i_a^- + v_b^- i_b^- + v_c^- i_c^-) \tag{5.28}$$

is returned to the power network, in other words the power loss

$$R_s[(i_a^-)^2 + (i_b^-)^2 + (i_c^-)^2]$$

is supplied by the unbalanced load which receives the positive-sequence power

$$v_A i_a^+ + v_B i_b^+ + v_C i_c^+ - R_s[(i_a^+)^2 + (i_b^+)^2 + (i_c^+)^2]$$

and outputs the power

$$v_A i_a^+ + v_B i_b^+ + v_C i_c^+ - R_s[(i_a^+)^2 + (i_b^+)^2 + (i_c^+)^2] - R_s[(i_a^-)^2 + (i_b^-)^2 + (i_c^-)^2$$

The active power flow sketched in Fig. 5.5 shows that the negative-sequence power does not contribute at all to the output power. This negative-sequence power is part of an energy conversion chain in which a small amount of positive-sequence power is converted into negative-sequence power that is supplied (returned) to the network.

If the main energy source is asymmetrical then some negative-sequence power supplied by the source is converted into output power.

A numerical example addressing a more general case that involves zero-sequence as well as inductive and capacitive impedances helps reinforce the understanding of the above phenomenon. The supply voltage (line-to-neutral) is

$$|v_s| = \sqrt{2}\,260 \begin{vmatrix} \sin(\omega t) \\ \sin(\omega t - 120°) \\ \sin(\omega t + 120°) \end{vmatrix} \quad \omega = 376.9 \text{ rad/s} = 21{,}600 \text{ deg/s}$$

An unbalanced load is supplied with the above voltage via a four-wire line, all conductors with $R_s = 0.1\ \Omega$ and $L_s = 2.65$ mH ($\omega L_s = 1.0\ \Omega$). The resulting measurements at the load terminals are as follows:

$$|\mathbf{V}| = \begin{vmatrix} \mathbf{V}_a \\ \mathbf{V}_b \\ \mathbf{V}_c \end{vmatrix} = \begin{vmatrix} 251\angle - 8° \\ 242\angle - 131° \\ 255\angle 122° \end{vmatrix} \text{V} \qquad |\mathbf{I}| = \begin{vmatrix} \mathbf{I}_a \\ \mathbf{I}_b \\ \mathbf{I}_c \\ \mathbf{I}_n \end{vmatrix} = \begin{vmatrix} 34.46\angle 5.54° \\ 46.08\angle - 152.2° \\ 10.55\angle 41.58° \\ 11.29\angle - 82.62° \end{vmatrix} \text{A}$$

The computed symmetrical components are

$$\begin{aligned} \mathbf{V}^+ &= 246.98 - j24.07 \text{ V} & \mathbf{I}^+ &= 25.14 - j10.52 \text{ A} \\ \mathbf{V}^- &= 16.68 \;\; - j10.42 & \mathbf{I}^- &= 8.69 \;\; - j17.57 \\ \mathbf{V}^0 &= -15.11 - j0.44 & \mathbf{I}^0 &= 0.48 \;\; - j3.72 \end{aligned}$$

The power generated is

$$P_{Gen} = 3 \times 260 \times \Re e\{\mathbf{I}^{+*}\} = 19607 \text{ W}$$

The power delivered by the load is

$$P_{out} = 3\Re e\{\mathbf{V}^+\mathbf{I}^{+*} + \mathbf{V}^-\mathbf{I}^{-*} + \mathbf{V}^0\mathbf{I}^{0*}\} = 19253 \text{ W}$$

The positive-sequence power supplied by the generator to the load is

$$P_{\substack{Suppl \\ Load}} = P_{Gen} - 3R_s(I^+)^2 = 19385 \text{ W}$$

Comparing the last two results we find that

$$P_{\substack{Suppl \\ Load}} - P_{out} = 19385 - 19253 = 132 \text{ W}$$

of positive-sequence power will not leave the electrical system, but will be converted into losses supplied by the negative- and the zero-sequence components.

Now the total power loss is

$$\Delta P = R_s(I_a^2 + I_b^2 + I_c^2 + I_n^2) = 354.96 \text{ W}$$

Each symmetrical component of the current contributes to the total power loss, i.e.

$$\Delta P = 3R_s[(I^+)^2 + (I^-)^2 + (I^0)^2] + R_s(3I^0)^2 = 3R_s[(I^+)^2 + (I^-)^2 + 4(I^0)^2] = 354.96 \text{ W}$$

The positive-sequence current causes the power loss

$$\Delta P^+ = 3R_s(I^+)^2 = 223.0 \text{ W}$$

The negative- and zero-sequence currents cause the power loss

$$\Delta P^- + \Delta P^0 = 3R_s[(I^+)^2 + 4(I^0)^2] = 131.26 \text{ W}$$

It is learned from here that the unbalanced load generates the energy that supports $\Delta P^- + \Delta P^0$, indeed

$$P^- = 3\Re e\{\mathbf{V}^-\mathbf{I}^{-*}\} = -114.55 \text{ W}$$
$$P^0 = 3\Re e\{\mathbf{V}^0\mathbf{I}^{0*}\} = -16.69 \text{ W}$$

and

$$|P^- + P^0| = \Delta P^- + \Delta P^0 = 131.24 \text{ W}$$

The same analysis is applied to reactive powers. The positive-sequence reactive power generated is

$$Q_{Gen} = 3 \times 260 \times \Im m\{\mathbf{I}^{+*}\} = 8207 \text{ var}$$

out of which

$$Q_{\substack{Suppl \\ Load}} = Q_{Gen} - 3\omega L_s(I^+)^2 = 5979.3 \text{ var}$$

is supplied to the load.

However, the reactive power measured at the load terminals is

$$Q = 3\Im m\{\mathbf{V}^+\mathbf{I}^{+*} + \mathbf{V}^-\mathbf{I}^{-*} + \mathbf{V}^0\mathbf{I}^{0*}\} = 4660 \text{ var}$$

We find that the difference between supplied and what can be called the actual reactive power of the load (the measured quantity) is $Q_{\substack{Suppl \\ Load}} - Q = 1319.3$ var. The negative- and the zero-sequence reactive powers are

$$Q^- = 3\Im m\{\mathbf{V}^-\mathbf{I}^{-*}\} = -1151 \text{ var}$$

$$Q^0 = 3\Im m\{\mathbf{V}^0\mathbf{I}^{0*}\} = -169 \text{ var}$$

and

$$Q^- + Q^0 = -1320 \text{ var}$$

leading to the following conclusion: Part of Q^+ is converted in Q^- and Q^0, and their instantaneous powers oscillate between the load and the line inductances. The Q^+ related oscillations take place between the main source and the line inductances as well as the load.

If the network has multiple loads a fraction of the P^- and P^0 flows also into some loads, where it is dissipated (motors) or converted (incandescent lamps and certain rectifiers). One may observe dominant unbalanced loads that generate non positive-sequence energy and smaller loads, balanced or not, that absorb such energy.

5.4.2   Expressing the Powers by Means of Symmetrical Components

In Fig. 5.6 is sketched the equivalent circuit of a three-phase, four-wire system where the asymmetrical line-to-neutral voltages are replaced with their symmetrical components, i.e.

$$
\begin{aligned}
v_a &= v_a^+ + v_a^- + v_a^0 \quad &\text{and} \quad i_a &= i_a^+ + i_a^- + i_a^0 \\
v_b &= v_b^+ + v_b^- + v_b^0 \quad &\text{and} \quad i_b &= i_b^+ + i_b^- + i_b^0 \\
v_c &= v_c^+ + v_c^- + v_c^0 \quad &\text{and} \quad i_c &= i_c^+ + i_c^- + i_c^0
\end{aligned} \tag{5.29}
$$

The instantaneous positive-sequence voltages and currents are:

$$
\begin{aligned}
v_a^+ &= \widehat{V}^+ \sin(\omega t + \alpha^+) \quad &\text{and} \quad i_a^+ &= \widehat{I}^+ \sin(\omega t + \alpha^+ - \theta^+) \\
v_b^+ &= \widehat{V}^+ \sin(\omega t + \alpha^+ - 120°) \quad &\text{and} \quad i_b^+ &= \widehat{I}^+ \sin(\omega t + \alpha^+ - \theta^+ - 120°) \\
v_c^+ &= \widehat{V}^+ \sin(\omega t + \alpha^+ + 120°) \quad &\text{and} \quad i_c^+ &= \widehat{I}^+ \sin(\omega t + \alpha^+ - \theta^+ + 120°)
\end{aligned} \tag{5.30}
$$

The instantaneous negative-sequence voltages and currents are:

$$
\begin{aligned}
v_a^- &= \widehat{V}^- \sin(\omega t + \alpha^-) \quad &\text{and} \quad i_a^- &= \widehat{I}^- \sin(\omega t + \alpha^- - \theta^-) \\
v_b^- &= \widehat{V}^- \sin(\omega t + \alpha^- + 120°) \quad &\text{and} \quad i_b^- &= \widehat{I}^- \sin(\omega t + \alpha^- - \theta^- + 120°) \\
v_c^- &= \widehat{V}^- \sin(\omega t + \alpha^- 120°) \quad &\text{and} \quad i_c^- &= \widehat{I}^- \sin(\omega t + \alpha^- - \theta^- 120°)
\end{aligned} \tag{5.31}
$$

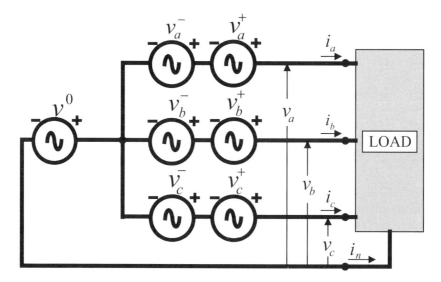

Figure 5.6   Equivalent three-phase, four-wire system with symmetrical components.

and the zero-sequence voltages and currents are:

$$v_a^0 = v_b^0 = v_c^0 = \widehat{V}^0 \sin(\omega t + \alpha^0) \quad \text{and} \quad i_a^0 = i_b^0 = i_c^0 = \widehat{I}^0 \sin(\omega t + \alpha^0 - \theta^0) \qquad (5.32)$$

The instantaneous power has nine terms [10]

$$p = v_a i_a + v_b i_b + v_c i_c = p^+ + p^- + p^0 + p^{+-} + p^{+0} + p^{-+} + p^{-0} + p^{0+} + p^{0-} \qquad (5.33)$$

The positive-sequence instantaneous power

$$p^+ = v_a^+ i_a^+ + v_b^+ i_b^+ + v_c^+ i_c^+ = p_p^+ + p_q^+ \qquad (5.34)$$

is composed of two terms: the *instantaneous positive-sequence active power*

$$p_p^+ = V^+ I^+ \cos\theta^+ \{3 - [\cos(2\omega t + 2\alpha^+) + \cos(2\omega t + 2\alpha^+ - 240°)$$
$$+ \cos(2\omega t + 2\alpha^+ + 240°)]\} = 3V^+ I^+ \cos(\theta^+) \qquad (5.35)$$

has an expression that shows the intrinsic powers cancellation. The second term is the *instantaneous positive-sequence reactive power*

$$p_q^+ = -V^+ I^+ \sin(\theta^+)[\sin(2\omega t + 2\alpha^+) +$$
$$+ \sin(2\omega t + 2\alpha^+ - 240°) + \sin(2\omega t + 2\alpha^+ + 240°)] = 0 \qquad (5.36)$$

Here we are met by the same question we were faced with when evaluating (5.3), if $p_q^+ = 0$, is there reactive power? And the answer is the same: there is a total positive-sequence reactive power $Q^+ = 3V^+ I^+ \sin(\theta^+)$ and each phase carries its own oscillations with the amplitude $Q^+/3$ supported by the Poynting vectors surrounding each conductor. This means that the total Poynting vector flux caused by the interaction between the positive-sequence electric field with the positive-sequence magnetic field caused by the reactive currents, through a close surface that envelopes the three-phase load, is nil, (see section 5.6).

For the negative- and the zero-sequence instantaneous powers we find similar expressions: the negative-sequence instantaneous power

$$p^- = v_a^- i_a^- + v_b^- i_b^- + v_c^- i_c^- = p_p^- + p_q^- \qquad (5.37)$$

with the *instantaneous negative-sequence active power*

$$p_p^- = V^- I^- \cos(\theta^-)\{3 - [\cos(2\omega t + 2\alpha^-) + \cos(2\omega t + 2\alpha^- + 240°)$$
$$+ \cos(2\omega t + 2\alpha^- - 240°)]\} = 3V^- I^- \cos(\theta^-) \qquad (5.38)$$

and the *instantaneous negative-sequence reactive power*

$$p_q^- = -V^- I^- \sin(\theta^-)[\sin(2\omega t + 2\alpha^-)$$
$$+ \sin(2\omega t + 2\alpha^- + 240°) + \sin(2\omega t + 2\alpha^- - 240°)] = 0 \qquad (5.39)$$

that characterizes the negative-sequence active and reactive powers, $P^- = 3V^-I^- \cos(\theta^-)$ and $Q = 3V^-I^- \sin(\theta^-)$, respectively.

The zero-sequence power has simpler expressions:

$$p^0 = v_a^0 i_a^0 + v_b^0 i_b^0 + v_c^0 i_c^0 = p_p^0 + p_q^0 \tag{5.40}$$

where

$$p_p^0 = 3V^0 I^0 \cos(\theta^0)[1 - (\cos(2\omega t + 2\alpha^0)] \tag{5.41}$$

$$p_q^0 = -3V^0 I^0 \sin(\theta^0) \sin(2\omega t + 2\alpha^0) \tag{5.42}$$

We observe that

$$p_p^+ + p_p^- + p_p^0 = 3V^+I^+ \cos(\theta^+) + 3V^-I^- \cos(\theta^-) + 3V^0 I^0 \cos(\theta^0) = P^+ + P^- + P^0 = P \tag{5.43}$$

The remaining six instantaneous elementary powers of (5.33) are nonactive, for example the positive-negative instantaneous power is

$$
\begin{aligned}
p^{+-} &= v_a^+ i_a^- + v_b^+ i_b^- + v_c^+ i_c^- \\
&= V^+ I^- [\cos(\alpha^+ - \alpha^- + \theta^-) + \cos(\alpha^+ - \alpha^- + \theta^- - 240°) \\
&\quad + \cos(\alpha^+ - \alpha^- + \theta^- + 240°) - 3\cos(2\omega t + \alpha^+ + \alpha^- - \theta^-)] \tag{5.44}
\end{aligned}
$$

At any moment the mean value of p^{+-} is zero, however, this instantaneous power has some remarkable peculiarity; each phase has an average power that flows unidirectionally. On phase a we have $P_a^{+-} = V^+I^- \cos(\gamma)$, on phase b there is $P_b^{+-} = V^+I^- \cos(\gamma - 240°)$, and on phase c we find $P_c^{+-} = V^+I^- \cos(\gamma + 240°)$, where $\gamma = \alpha^+ - \alpha^- + \theta^-$.

The total Poynting vector flux caused by the interaction between the positive-sequence electric field with the negative-sequence magnetic field through a close surface that envelopes the three-phase load is nil. The peculiarity of this nonactive energy flow is that along one of the conductors the Poynting vector is "pumping" energy toward the load and along the other two conductors this very energy is returned.

The same observations apply to the remaining instantaneous powers; in steady-state each of the powers p^{+0}, p^{-+}, p^{-0}, p^{0+}, and p^{0-} has zero average power and each one of these powers has a component flowing continuously to or from the load on one phase and returning via the other two phases. This is in agreement with the fact that unbalanced conditions lead to different power loss values among the phases.

Now let us return to the four-conductors line with $R_a = R_b = R_c = R_s$ and $R_n \neq R_s$. From (5.22) results that the effective current squared is

$$I_e^2 = (I_a^2 + I_b^2 + I_c^2 + \rho I_n^2)/3 \, ; \quad \rho = R_n/R_s \tag{5.45}$$

Each one of the rms currents squared of (5.45) can be expressed in function of symmetrical components:

$$
\begin{aligned}
I_a^2 &= \mathbf{I}_a \mathbf{I}_a^* = (\mathbf{I}^+ + \mathbf{I}^- + \mathbf{I}^0)(\mathbf{I}^{+*} + \mathbf{I}^{-*} + \mathbf{I}^{0*}) \\
&= (I^+)^2 + (I^-)^2 + (I^0)^2 + \mathbf{I}^+\mathbf{I}^{-*} + \mathbf{I}^+\mathbf{I}^{0*} + \mathbf{I}^-\mathbf{I}^{+*} + \mathbf{I}^-\mathbf{I}^{0*} + \mathbf{I}^0\mathbf{I}^{+*} + \mathbf{I}^0\mathbf{I}^{-*} \\
I_b^2 &= \mathbf{I}_b \mathbf{I}_b^* = (\mathbf{I}^+ + \mathbf{a}^2\mathbf{I}^- + \mathbf{a}\mathbf{I}^0)(\mathbf{I}^{+*} + \mathbf{a}\mathbf{I}^{-*} + \mathbf{a}^2\mathbf{I}^{0*}) \quad \mathbf{a} = 1\angle 120^\circ \\
&= (I^+)^2 + (I^-)^2 + (I^0)^2 + \mathbf{a}\mathbf{I}^+\mathbf{I}^{-*} + \mathbf{a}^2\mathbf{I}^+\mathbf{I}^{0*} + \mathbf{a}^2\mathbf{I}^-\mathbf{I}^{+*} + \mathbf{a}\mathbf{I}^-\mathbf{I}^{0*} + \mathbf{a}\mathbf{I}^0\mathbf{I}^{+*} + \mathbf{a}^2\mathbf{I}^0\mathbf{I}^{-*} \\
I_c^2 &= \mathbf{I}_c \mathbf{I}_c^* = (\mathbf{I}^+ + \mathbf{a}\mathbf{I}^- + \mathbf{a}^2\mathbf{I}^0)(\mathbf{I}^{+*} + \mathbf{a}^2\mathbf{I}^{-*} + \mathbf{a}\mathbf{I}^{0*}) \\
&= (I^+)^2 + (I^-)^2 + (I^0)^2 + \mathbf{a}^2\mathbf{I}^+\mathbf{I}^{-*} + \mathbf{a}\mathbf{I}^+\mathbf{I}^{0*} + \mathbf{a}\mathbf{I}^-\mathbf{I}^{+*} + \mathbf{a}^2\mathbf{I}^-\mathbf{I}^{0*} + \mathbf{a}^2\mathbf{I}^0\mathbf{I}^{+*} + \mathbf{a}\mathbf{I}^0\mathbf{I}^{-*} \\
I_n^2 &= \mathbf{I}_n \mathbf{I}_n^* = 3\mathbf{I}^0 3\mathbf{I}^{0*} = 9(I^0)^2
\end{aligned}
\tag{5.46}
$$

Substitution of (5.46) in (5.45) gives

$$
I_e^2 = (I^+)^2 + (I^-)^2 + (1 + 3\rho)(I^0)^2
\tag{5.47}
$$

In a similar way is obtained the effective voltage squared

$$
\begin{aligned}
V_e^2 &= \frac{(V_{ab}^2 + V_{bc}^2 + V_{ca}^2)}{9} \\
&= \frac{(\mathbf{V}_a - \mathbf{V}_b)(\mathbf{V}_a^* - \mathbf{V}_b^*) + (\mathbf{V}_b - \mathbf{V}_c)(\mathbf{V}_b^* - \mathbf{V}_c^*) + (\mathbf{V}_c - \mathbf{V}_a)(\mathbf{V}_c^* - \mathbf{V}_a^*)}{9} \\
&= (V^+)^2 + (V^-)^2
\end{aligned}
\tag{5.48}
$$

Since the line-to-line voltages are used to define V_e, no zero-sequence voltage appears in (5.48). If the load is not Δ-connected the expression (5.48) is only an approximation. A detailed analysis of V_e is given in section 5.5.4.

The line power loss is

$$
\Delta P = R_s(I_a^2 + I_b^2 + I_c^2) + R_n I_n^2 = 3R_s I_e^2 = \frac{R_s}{3V_e^2}(3V_e I_e)^2 = \frac{R_s}{3V_e^2}S_e^2
\tag{5.49}
$$

Equation (5.49) proves that the Buchholz-Goodhue apparent power definition fulfills the much desired linear correlation between ΔP and the apparent power squared, S_e^2.

The ΔP expression written as a function of symmetrical components of the current is

$$
\Delta P = 3R_s[(I^+)^2 + (I^-)^2 + (1 + 3\rho)(I^0)^2]
\tag{5.50}
$$

The vector and the arithmetic apparent powers have the expressions

$$
\begin{aligned}
\mathbf{S}_V &= \mathbf{S}_a + \mathbf{S}_b + \mathbf{S}_c = \mathbf{V}_a \mathbf{I}_a^* + \mathbf{V}_b \mathbf{I}_b^* + \mathbf{V}_c \mathbf{I}_c^* \\
&= 3(\mathbf{V}^+\mathbf{I}^{+*} + \mathbf{V}^-\mathbf{I}^{-*} + \mathbf{V}^0\mathbf{I}^{0*})
\end{aligned}
\tag{5.51}
$$

and

$$
S_A = S_a + S_b + S_c = |\mathbf{V}_a \mathbf{I}_a^*| + |\mathbf{V}_b \mathbf{I}_b^*| + |\mathbf{V}_c \mathbf{I}_c^*|
\tag{5.52}
$$

When the graphs ΔP versus S_e^2, S_V^2, and S_e^2 are plotted for an unbalanced load situation it is found that only the graph ΔP versus S_e^2 is linear, (see problem 5.3).

The apparent power definition, $S_e = 3V_e I_e$, is the only known apparent power expression that fulfills the requirement $\Delta P = (R_S/3V_e^2)S_e^2$ and leads to a meaningful power factor. However, the effective apparent power as presented in the last two sections is based on the electrothermal effect and does not consider the nature of the load-motor, rectifier, arcing device, etc. More accurate definitions based on the actual energy conversion physical mechanism are needed.

In the following section are presented effective apparent power resolutions that have started to gain acceptance among researchers as well as engineers.

5.5 Effective Apparent Power Resolutions

5.5.1 FBD-Method

A few engineers understood the significant advantages of Buchholz's expression and advocated the separation of S_e in orthogonal components useful in the evaluation of the energy flow and operation of power systems [11,12,13]. Among all these methods the most salient place is given to the FBD-method, the Fryze-Buchholz-Depenbrock approach [14].

We consider a three-phase unbalanced load supplied with an asymmetrical three-phase voltage, Fig. 5.7a. The four currents obey the relation

$$i_A + i_B + i_C + i_N = 0 \tag{5.53}$$

The following analysis is based on an equivalent three-phase system where an artificial neutral point O is used as a reference point, Fig. 5.7.b. The artificial neutral point has the following essential properties:

1. The line to neutral voltages v_{AN}, v_{BN}, v_{CN} remain unchanged (same in Figs. 5.7a and 5.7.b).

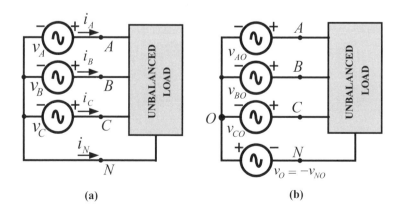

(a)   (b)

Figure 5.7   The FBD-method: (a) Actual circuit. (b) Equivalent circuit with neutral point O.

2. The node N is considered the terminal of a fourth phase.

3. The four line-to-the artificial neutral voltages obey the relation

$$v_{AO} + v_{BO} + v_{CO} + v_{NO} = 0 \qquad (5.54)$$

One may rewrite (5.54) using the notation $v_{ON} = v_O$:

$$v_{AN} - v_O + v_{BN} - v_O + v_{CN} - v_O - v_O = 0 \qquad (5.55)$$

Equation (5.55) gives the potential at the neutral point with respect to the node O:

$$v_{NO} = -v_O = -\frac{1}{4}(v_{AN} + v_{BN} + v_{CN}) \qquad (5.56)$$

and the voltages line-to-the artificial neutral:

$$v_{AO} = v_A - v_O = v_{AN} - \frac{1}{4}(v_{AN} + v_{BN} + v_{CN})$$

$$v_{BO} = v_B - v_O = v_{BN} - \frac{1}{4}(v_{AN} + v_{BN} + v_{CN})$$

$$v_{CO} = v_C - v_O = v_{CN} - \frac{1}{4}(v_{AN} + v_{BN} + v_{CN}) \qquad (5.57)$$

It also can be proved that the line-to-line load voltages remain unchanged, for example:

$$v_{AB} = v_{AO} - v_{BO} = (v_A - v_O) - (v_B - v_O) = v_A - v_B \qquad (5.58)$$

and this means that the line currents i_A, i_B, i_C, and i_N remain unchanged too. Moreover, the value of the instantaneous power p is not affected by v_O,

$$\begin{aligned} p &= v_{AO}i_A + v_{BO}i_B + v_{CO}i_C - v_O i_N \\ &= v_A i_A + v_B i_B + v_C i_C - v_O(i_A + i_B + i_C + i_N) \\ &= v_A i_A + v_B i_B + v_C i_C \end{aligned} \qquad (5.59)$$

Depenbrock analyzes the equivalent circuit shown in Fig. 5.7b as follows: each one of the four branch impedances is supplied with the apparent power,

$$\mathbf{S}_k = \mathbf{V}_k \mathbf{I}_k^* = P_k + j Q_k ; \quad k = \text{A, B, C, N}$$

The branch current i_k (the line current) is separated into two components: an in-phase component $i_\parallel$, (Depenbrock calls it *proportional current*) and an in-quadrature current $i_\perp$ (Depenbrock calls it *orthogonal current*). Below are the expressions of these currents:

$$i_{k\parallel} = G_k v_{kO} \quad \text{with} \quad G_k = \frac{P_k}{V_k^2} \qquad (5.60)$$

and

$$i_{k\perp} = B_k v_{kO} \quad \text{with} \quad B_k = \frac{-Q_k}{V_k^2} \tag{5.61}$$

This current division leads to a first representation of the branch shown in Fig. 5.8a where G_k and B_k are the branch k conductance and susceptance, respectively. Pivotal to this approach is a voltage v_Σ, that was introduced by Buchholz [5, 15]. Buchhollz called v_Σ and i_Σ collective quantities, *collective (instantaneous) voltage* and *current*. Their expressions are as follows:

$$v_\Sigma^2 = v_{AO}^2 + v_{BO}^2 + v_{CO}^2 + v_O^2 \tag{5.62}$$

$$i_\Sigma^2 = i_A^2 + i_B^2 + i_C^2 + i_N^2 \tag{5.63}$$

Thus leading straight to the *collective rms voltage* and *current*

$$V_\Sigma^2 = V_{AO}^2 + V_{BO}^2 + V_{CO}^2 + V_O^2 \tag{5.64}$$

$$I_\Sigma^2 = I_A^2 + I_B^2 + I_C^2 + I_N^2 \tag{5.65}$$

The system's apparent power is defined

$$S_\Sigma = V_\Sigma I_\Sigma \tag{5.66}$$

and it will be proved later that this is the maximum active power that can be transmitted to the unbalanced load by the asymmetrical voltages, under the condition that the rms collective voltage and current remain unchanged. This condition fulfills also Lyon's recommendation, (see section 5.3 and Appendix VI).

On the next step a *balanced conductance* or *active equivalent conductance* is defined using the collective active power and rms voltage:

$$G = \frac{P_\Sigma}{V_\Sigma^2} \; ; \quad P_\Sigma = \sum P_k \; ; \quad k = \text{A, B, C, N} \tag{5.67}$$

In the same way the *balanced equivalent susceptance* is defined

$$B = \frac{-Q_\Sigma}{V_\Sigma^2} \; ; \quad Q_\Sigma = \sum Q_k \; ; \quad k = \text{A, B, C, N} \tag{5.68}$$

The equivalent conductance G and susceptance B correspond to a balanced load. To separate the unbalanced components the *unbalanced branch conductances* ΔG_k and *susceptances* ΔB_k are defined:

$$\Delta G_k = G_k - G \quad \text{and} \quad \Delta B_k = B_k - B \tag{5.69}$$

The values of ΔG_k and ΔB_k may vary in magnitude from branch to branch.

An improved representation of the branch can be achieved by involving four currents, Fig. 5.8b: a *balanced current* $i_{kb} = i_{k\|b} + i_{k\perp b}$ and an *unbalanced current* $i_{ku} = i_{k\|u} + i_{k\perp u}$. The

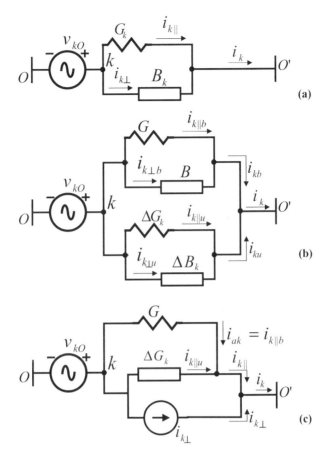

Figure 5.8 Equivalent branch circuits using the FBD-method: (a) Basic approach; proportional and orthogonal currents. (b) Use of balanced and unbalanced currents. (c) Use of active, nonactive, and proportional unbalanced current. Source: M. Depenbrock, "The FBD-Method, A Generally Applicable tool for Analyzing Power Relations," IEEE Trans. On Power Systems, Vol.8, No.2, May 1993, pp.381–87.

rms values and names of these four currents are as follows:

$$I_{k\|b} = G \, V_{Ok} \tag{5.70}$$

is the *proportional branch current,*

$$I_{k\perp b} = |B| \, V_{Ok} \tag{5.71}$$

is the *orthogonal branch current,*

$$I_{k\|u} = |\Delta G_k| \, V_{Ok} \tag{5.72}$$

is the *proportional unbalance branch current,*

$$I_{k\perp u} = |\Delta B_k| \, V_{Ok} \tag{5.73}$$

is the *orthogonal unbalance branch current*. These four currents are mutually orthogonal and lead to the following total collective rms current expression

$$I_\Sigma^2 = \frac{1}{T} \int_0^T (i_A^2 + i_B^2 + i_C^2 + i_N^2) \, dt = \frac{1}{T} \int_0^T (i_{\Sigma\|b}^2 + i_{\Sigma\perp b}^2 + i_{\Sigma\|u}^2 + i_{\Sigma\perp u}^2) \, dt$$
$$= I_{\Sigma\|b}^2 + I_{\Sigma\perp b}^2 + I_{\Sigma\|u}^2 + I_{\Sigma\perp u}^2 \tag{5.74}$$

Multiplying (5.74) with V_Σ^2, each set of collective currents leads to a component of the system apparent power

$$S_\Sigma^2 = V_\Sigma^2 I_\Sigma^2 = P_\Sigma^2 + F_\Sigma^2 = P_\Sigma^2 + Q_\Sigma^2 + F_{\Sigma\|u}^2$$
$$= P_\Sigma^2 + Q_\Sigma^2 + F_{\Sigma\|u}^2 + F_{\Sigma\perp u}^2 \tag{5.75}$$

where

$$P_\Sigma = V_\Sigma I_{\Sigma\|b}$$

is the *active power*, and

$$F_\Sigma = \sqrt{Q_\Sigma^2 + F_{\Sigma u}^2}$$

is the *total nonactive power*, that can be separated into *reactive power*:

$$Q_\Sigma = V_\Sigma I_{\Sigma\perp b}$$

and the *unbalanced nonactive power*:

$$F_{\Sigma u} = \sqrt{F_{\Sigma\|u}^2 + F_{\Sigma\perp u}^2}$$

where

$$F_{\Sigma\|u} = V_\Sigma I_{\Sigma\|u}$$

is the *proportional unbalance nonactive power*, and

$$F_{\Sigma\perp u} = V_\Sigma I_{\Sigma\perp u}$$

is the *orthogonal nonactive power.*

The nature of each one of these three nonactive powers differs: The reactive power Q_Σ is due to energy that oscillates between source and energy storage components, while the powers $F_{\Sigma\|u}$ and $F_{\Sigma\perp u}$ are due only to load unbalance.

There are many ways in which the branch equivalent circuit can be further solved. A simple approach, close to Fryze's concept and adopted by the German standard [16] DIN 40110 is sketched in Fig. 5.8c where all the orthogonal components and the reactive power are lumped together and modeled by the current source $i_{k\perp}$. Depenbrock resolves this model using only three components:

$$S_\Sigma^2 = P_\Sigma^2 + Q_{tot_\Sigma\|}^2 + Q_{tot_\Sigma\perp}^2 \tag{5.76}$$

where

$$Q_{tot_\Sigma\|} = V_\Sigma I_{\Sigma\|u} \quad \text{and} \quad Q_{tot_\Sigma\perp} = V_\Sigma I_{\Sigma\perp} \; ; \quad I_{\Sigma\perp} = \sqrt{I_{\Sigma\perp b}^2 + I_{\Sigma\perp u}^2}$$

The complete three-phase, four-wire equivalent circuit derived from [16, 25] is given in Fig. 5.9.

A more complex, but most interesting, model that makes use of symmetrical components v^+, v^-, and v^0, is shown in Fig. 5.10. Based on (5.57) the branch voltages are separated into symmetrical components as follows:

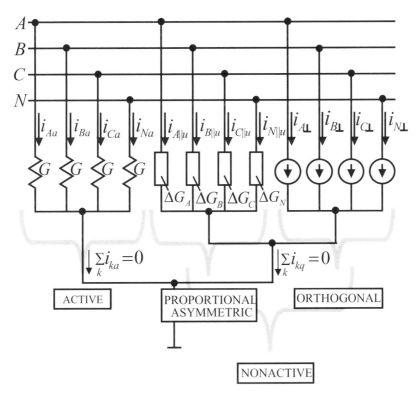

Figure 5.9 Three-phase, four-wire load: Depenbrock's equivalent circuit [16, 25].

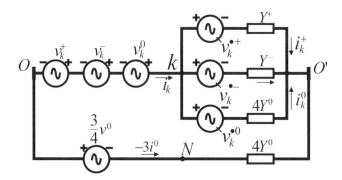

Figure 5.10   Depenbrock's equivalent branch circuit using symmetrical components.

$$v_{A0} = v_A - v_0 = v_A^+ + v_A^- + v_A^0 - \frac{1}{4}(v_A + v_B + v_C) = v_A^+ + v_A^- + \frac{1}{4}v^0$$

$$v_{B0} = v_B^+ + v_B^- + \frac{1}{4}v^0$$

$$v_{C0} = v_C^+ + v_C^- + \frac{1}{4}v^0$$

$$v_{N0} = \frac{-3}{4}v^0 \qquad v_A^0 = v_B^0 = v_C^0 = v^0 \tag{5.77}$$

The symmetrical components of the current, I_A^+, I_A^-, and I^0 are handled as follows: Positive-, negative- and zero-sequence currents flow in each branch; in turn each one of the four branches is subdivided into three parallel subbranches such that each one of them is dedicated to one of the sequence currents. The subbranches have the admittances

$$Y^+ = \frac{I_A^+}{V_A^+} = \frac{I_B^+}{V_B^+} = \frac{I_C^+}{V_C^+}$$

$$Y^- = \frac{I_A^-}{V_A^-} = \frac{I_B^-}{V_B^-} = \frac{I_C^-}{V_C^-}$$

and

$$Y^0 = \frac{I^0}{4V^0}$$

Each admittance is connected in series with a complementary voltage

$$Y^+ \quad \text{with} \quad v_k^{\bullet+} = v_k^- + v^0$$
$$Y^- \quad \text{with} \quad v_k^{\bullet-} = v_k^+ + v^0$$

and

$$Y^0 \quad \text{with} \quad v_k^{\bullet 0} = v_k^+ + v^-$$

The complementary voltage is needed to ensure that the voltage impressed across the admittance Y^+ is v^+, across Y^- is v^-, and across Y^0 is v^0.

This time each branch has six distinctive powers that lead to the S_Σ resolution with six components:

$$S_\Sigma^2 = (S_\Sigma^+)^2 + (F_\Sigma^{\bullet+})^2 + (S_\Sigma^-)^2 + (F_\Sigma^{\bullet-})^2 + (S_\Sigma^0)^2 + (F_\Sigma^{\bullet0})^2 \qquad (5.78)$$

where $S_\Sigma^+ = 3V^+I^+$, $S_\Sigma^- = 3V^-I^-$, and $S_\Sigma^0 = 3V^0I^0$ are the classical symmetrical components powers and $F_\Sigma^{\bullet+} = 3V_\Sigma^{\bullet+}I^+$, $F_\Sigma^{\bullet-} = 3V_\Sigma^{\bullet-}I^-$ and $F_\Sigma^{\bullet0} = 3V_\Sigma^{\bullet0}I^0$ are complementary powers, all three nonactive powers caused by the fact that the system is not balanced.

The admittances Y^+, Y^-, and Y^0 can in turn be described by more complex subcircuits. If one chooses the Fig. 5.10 approach then, similarly to (5.75), we can write:

$$
\begin{aligned}
(S_\Sigma^+)^2 &= (P_\Sigma^+)^2 + (Q_\Sigma^+)^2 + (F_{\Sigma u}^+)^2 + (F_{\Sigma \perp u}^+)^2 \\
(S_\Sigma^-)^2 &= (P_\Sigma^-)^2 + (Q_\Sigma^-)^2 + (F_{\Sigma u}^-)^2 + (F_{\Sigma \perp u}^-)^2 \\
(S_\Sigma^0)^2 &= (P_\Sigma^0)^2 + (Q_\Sigma^0)^2 + (F_{\Sigma u}^0)^2 + (F_{\Sigma \perp u}^0)^2
\end{aligned}
\qquad (5.79)
$$

5.5.2   L. S. Czarnecki's Method

A simplified approach based on Buchholz's and Fryze's theories and similar to Depenbrock's was introduced in 1988 [17]. The active and reactive powers of the load are obtained from the basic expressions

$$
\begin{aligned}
P &= \Re e\{\mathbf{V}_a\mathbf{I}_a^* + \mathbf{V}_b\mathbf{I}_b^* + \mathbf{V}_c\mathbf{I}_c^*\} \\
Q &= \Im m\{\mathbf{V}_a\mathbf{I}_a^* + \mathbf{V}_b\mathbf{I}_b^* + \mathbf{V}_c\mathbf{I}_c^*\}
\end{aligned}
$$

An equivalent conductance $G_e = P/V_\Sigma^2$ and an equivalent susceptance $B_e = -Q/V_\Sigma^2$ (V_Σ is Buchholz's collective rms voltage), help define the active currents

$$
\begin{aligned}
i_{Aa} &= G_e v_A \\
i_{Ba} &= G_e v_B \\
i_{Ca} &= G_e v_C
\end{aligned}
\qquad (5.80)
$$

and the reactive currents

$$
\begin{aligned}
i_{Ar} &= B_e \frac{d}{d(\omega t)} v_A = \sqrt{2}\Re e\{_JB_e\mathbf{V}_A\epsilon^{J\omega t}\} \\
i_{Br} &= B_e \frac{d}{d(\omega t)} v_B = \sqrt{2}\Re e\{_JB_e\mathbf{V}_B\epsilon^{J\omega t}\} \\
i_{Cr} &= B_e \frac{d}{d(\omega t)} v_C = \sqrt{2}\Re e\{_JB_e\mathbf{V}_C\epsilon^{J\omega t}\}
\end{aligned}
\qquad (5.81)
$$

The line currents consist of three terms

$$i_A = i_{Aa} + i_{Ar} + i_{Au}$$
$$i_B = i_{Ba} + i_{Br} + i_{Bu}$$
$$i_C = i_{Ca} + i_{Cr} + i_{Cu} \qquad (5.82)$$

where the currents i_{Au}, i_{Bu} and i_{Cu} are due to load unbalance.

Czarnecki proved that the three collective currents i_{a_Σ}, i_{r_Σ}, and i_{u_Σ} are mutually orthogonal. He used the following elegant demonstration based on the mathematical identity:

$$\frac{1}{T} \int_0^T f_i(t) f_j(t) \, dt = \Re e\{\mathbf{F}_i \mathbf{F}_j^*\}$$

where $f_i(t)$ and $f_j(t)$ are sinusoidal functions with identical frequencies and $\mathbf{F}_i$ and $\mathbf{F}_i$ are their respective rms phasors.

The total collective rms current squared is

$$\begin{aligned}
I_\Sigma^2 &= \frac{1}{T} \int_0^T (i_A^2 + i_B^2 + i_C^2) \, dt \\
&= \frac{1}{T} \int_0^T [(i_{Aa} + i_{Ar} + i_{Au})^2 + (i_{Ba} + i_{Br} + i_{Bu})^2 + (i_{Ca} + i_{Cr} + i_{Cu})^2] \, dt \\
&= \frac{1}{T} \int_0^T [i_{a_\Sigma}^2 + i_{r_\Sigma}^2 + i_{u_\Sigma}^2] \, dt + 2[\text{Ц}(ar) + \text{Ц}(ur) + \text{Ц}(ua))] \qquad (5.83)
\end{aligned}$$

where

$$\begin{aligned}
\text{Ц}(ar) &= \frac{1}{T} \int_0^T (i_{Aa} i_{Ar} + i_{Ba} i_{Br} + i_{Ca} i_{Cr}) \, dt \\
&= \Re e\{{}_J B_e \mathbf{V}_A G_e \mathbf{V}_A^* + {}_J B_e \mathbf{V}_B G_e \mathbf{V}_B^* + {}_J B_e \mathbf{V}_C G_e \mathbf{V}_C^*\} \\
&= \Re e\{{}_J B_e G_e (V_A^2 + V_B^2 + V_C^2)\} = 0 \qquad (5.84)
\end{aligned}$$

$$\begin{aligned}
\text{Ц}(ur) &= \frac{1}{T} \int_0^T (i_{Au} i_{Ar} + i_{Bu} i_{Br} + i_{Cu} i_{Cr}) \, dt \\
&= \frac{1}{T} \int_0^T [(i_A - i_{Aa} - i_{Ar}) i_{Ar} + (i_B - i_{Ba} - i_{Br}) i_{Br} + (i_C - i_{Ca} - i_{Cr}) i_{Cr}] \, dt \\
&= \frac{1}{T} \int_0^T [(i_{Ar} i_A - i_{Aa} i_{Ar} - i_{Ar}^2) + (i_{Br} i_B - i_{Ba} i_{Br} - i_{Br}^2) \\
&\quad + (i_{Cr} i_C - i_{Ca} i_{Cr} - i_{Cr}^2)] \, dt \\
&= \Re e\{{}_J B_e \mathbf{V}_A \mathbf{I}_A^* + {}_J B_e \mathbf{V}_B \mathbf{I}_B^* + {}_J B_e \mathbf{V}_C \mathbf{I}_C^*\} - B_e^2 V_\Sigma^2 \\
&= B_e \, \Re e\{{}_J (P + {}_J Q)\} + B_e Q = -B_e Q + B_e Q = 0 \qquad (5.85)
\end{aligned}$$

$$
\begin{aligned}
Ц(ua) &= \frac{1}{T}\int_0^T (i_{Au}i_{Aa} + i_{Bu}i_{Ba} + i_{Cu}i_{Ca})\,dt = \frac{1}{T}\int_0^T [(i_A - i_{Aa} - i_{Ar})i_{Ara} \\
&\quad + (i_B - i_{Ba} - i_{Br})i_{Ba} + (i_C - i_{Ca} - i_{Cr})i_{Ca}]\,dt \\
&= \frac{1}{T}\int_0^T [(i_{Aa}i_A - i_{Aa}i_{Ar} - i_{Aa}^2) + (i_{Ba}i_B - i_{Ba}i_{Br} - i_{Ba}^2) \\
&\quad + (i_{Ca}i_C - i_{Ca}i_{Cr} - i_{Ca}^2)]\,dt \\
&= \frac{1}{T}[G_e(v_A i_A + v_B i_B + v_C i_C) - G_e^2(v_A^2 + v_B^2 + v_C^2) \\
&\quad - I_{Aa}i_{Ar} - I_{Ba}i_{Br} - I_{Ca}i_{Cr}]\,dt \\
&= G_e P - G_e^2 V_\Sigma^2 = G_e(P - P) = 0
\end{aligned}
\tag{5.86}
$$

thus from (5.83) results

$$
I_\Sigma^2 = I_{a_\Sigma}^2 + I_{r_\Sigma}^2 + I_{u_\Sigma}^2
\tag{5.87}
$$

and

$$
S_\Sigma = V_\Sigma I_\Sigma = \sqrt{P^2 + Q^2 + D_u^2}
\tag{5.88}
$$

We recognize that (5.88) is identical to (5.75) where $D_u^2 = F_{\Sigma u}^2 = F_{\Sigma\parallel u}^2 + F_{\Sigma\perp u}^2$. Moreover, this model works well only for three-phase three-wire systems.

5.5.3 IEEE Std. 1459–2010 Method

This method adheres to Lyon's concept of apparent power with an additional requirement: for the ideal condition used to supply the maximum power (5.18), represented by the apparent power $S_e = 3V_e I_e$, the hypothetical three-phase compensated system supplies a perfectly balanced and purely resistive load with perfectly symmetrical line currents.

This method is the only one known to rely on an apparent power definition that satisfies the key condition

$$
\Delta P = \frac{R_s}{V_{LL}^2} S_e^2
$$

The effective current I_e, is the positive-sequence current conceived by Buchholz, Lyon and Goodhue. Its mathematical expression (5.22) considers the actual differences between the neutral wire (or the neutral current return path) resistance R_n and the line resistance R_s (or resistances R_a, R_b, R_c, that are not necessarily equal (5.20))

$$
I_e = \sqrt{\frac{I_a^2 + I_b^2 + I_c^2 + \rho I_n^2}{3}} = \sqrt{(I^+)^2 + (I^-)^2 + (1 + 3\rho)(I^0)^2}
$$

where the ratio $\rho = R_n/R_s$. If the value of ρ is not known the IEEE Std. 1459–2010 recommends using $\rho = 1.0$. The FBD-method does not use the ratio ρ, nevertheless checking the definition of I_Σ (5.65) one can recognize that implicitly it corresponds to $\rho = 1$ in which case $\sqrt{3}\, I_e = I_\Sigma$.

The ratio ρ affects significantly the values of I_e and S_e. If one uses $\rho = 1.0$ for the actual measurement, when the true $\rho > 1.0$, the expression

$$100\left[\frac{S_{e(\rho\neq1)} - S_{e(\rho=1)}}{S_{e(\rho\neq1)}}\right] = 100\left[\frac{I_{e(\rho\neq1)} - I_{e(\rho=1)}}{I_{e(\rho\neq1)}}\right] = 100\left[1 - \sqrt{\frac{(I^-/I^+)^2 + 4(I^0/I^+)^2 + 1}{(I^-/I^+)^2 + (3+\rho)(I^0/I^+)^2 + 1}}\,\right]$$

helps compute the relative difference between the two measurements. The graphs for the most unfavorable case, when $I^-/I^+ = 0$, are summarized in Fig. 5.11 for I^0/I^+ parameter. One learns that for $\rho < 1.0$ the measured apparent power with $\rho = 1$ is larger than the actual, and the trend reverses for $\rho > 1.0$. In actual distribution systems $\rho > 1$ and assuming $\rho = 1$ will favor the consumers [23] (see Appendix VII).

For a three-wire system $I^0 = 0$ and

$$I_e = \sqrt{\frac{I_a^2 + I_b^2 + I_c^2}{3}} = \sqrt{(I^+)^2 + (I^-)^2}$$

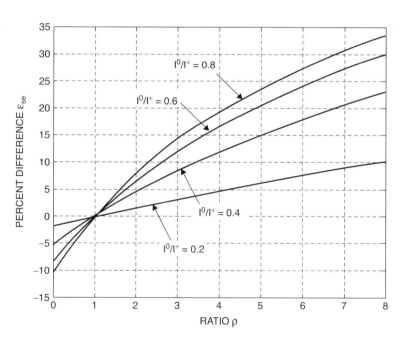

Figure 5.11 Percent apparent power difference versus ρ. $100(S_{e(\rho\neq1)} - S_{e(\rho=1)})/S_{e(\rho\neq1)}$ for $I^-/I^+ = 0$ [23]. Source: S. Pajic, A. E. Emanuel, "Effect of Neutral Path Power losses on the Apparent Power Definition: a Preliminary Study," IEEE Trans. On Power Delivery, Vol.24, No.2, April 2009, pp.517–523.

The expression of the effective voltage V_e is discussed in detail in the next section (5.109).

The effective apparent power S_e can be conveniently separated on its main component S^+, the positive-sequence apparent power and the remaining term S_u, the *unbalance apparent power*:

$$S_u^2 = S_e^2 - (S^+)^2 \tag{5.89}$$

with

$$(S^+)^2 = (P^+)^2 + (Q^+)^2 \tag{5.90}$$

This approach emphasizes the positive-sequence components as the most significant powers and considers also a positive-sequence power factor,

$$PF^+ = \frac{P^+}{S^+} \tag{5.91}$$

Under sinusoidal conditions, the resolution of the apparent power with lumped active and reactive powers ($P = P^+ + P^- + P^0$ and $Q = Q^+ + Q^- + Q^0$) is another possibility that leads to an expression similar to (5.75)

$$S_e^2 = (3V_e I_e)^2 = P^2 + Q^2 + N_u^2 \qquad N_u \approx F_{\Sigma u}^2 \tag{5.92}$$

More details on the IEEE Std. 1459–2010 are presented in the next section.

5.5.4 Comparison Between The Two Major Engineering Schools of Thought

The FBD-method was embraced by many scholars as the theory that leads to the ultimate definition of apparent power that equals the maximum active power that can be transmitted under given constraints [18, 19, 20].

Following is the basic explanation that presents the concept of unity power factor according to the FBD-method:

The unbalanced load shown in Fig. 5.12a is compensated as illustrated in Fig. 5.12b. The equivalent circuit of the load plus compensator corresponds to a set of four equal resistances R_Σ connected to a virtual neutral point O, Fig. 5.12c.

The resistance R_Σ has the value

$$R_\Sigma = \frac{V_{AO}}{I_{AO}} = \frac{V_{BO}}{I_{BO}} = \frac{V_{CO}}{I_{AO}} = \frac{V_{NO}}{I_{AO}} \tag{5.93}$$

and combined with the constraint

$$I_A^2 + I_B^2 + I_C^2 + I_N^2 = I_{AO}^2 + I_{BO}^2 + I_{CO}^2 + I_{NO}^2 \tag{5.94}$$

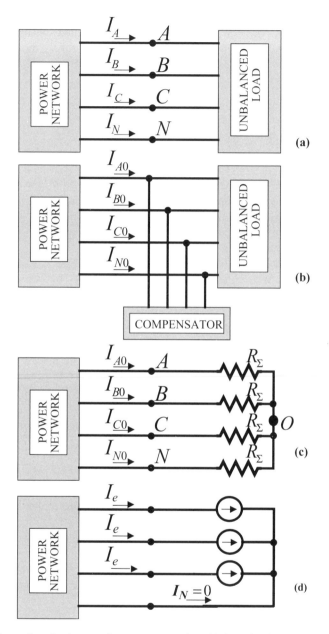

Figure 5.12   Three-phase load power factor compensation: (a) Actual unbalanced three-phase system. (b) Compensated system with line power loss unchanged (additional unity power factor load not shown). (c) Equivalent compensated load, FBD-method. (d) Equivalent compensated load, IEEE Std. 1459–2010. (Positive-sequence currents in-phase with fundamental).

results in

$$R_\Sigma^2 = \frac{V_\Sigma^2}{I_\Sigma^2} \tag{5.95}$$

where

$$V_\Sigma^2 = V_{AO}^2 + V_{AO}^2 + V_{AO}^2 + V_{AO}^2 \tag{5.96}$$

and

$$I_\Sigma^2 = I_{AO}^2 + I_{BO}^2 + I_{CO}^2 + I_{NO}^2 = I_A^2 + I_B^2 + I_C^2 + I_N^2 \tag{5.97}$$

are Buchholz's collective rms voltage and current squared, the key electrical values to the S_Σ definition.

Equation (5.94) implies that the four conductors have equal resistances and the compensated system has the same line power loss as the original system, i.e., the neutral current path resistance R_N equals the line resistance R_s. Obviously this condition is not always realistic.

The power delivered to the hypothetical system equivalent to four resistances R_Σ is

$$P_{max} = S_\Sigma = V_{AO}I_{AO} + V_{BO}I_{BO} + V_{CO}I_{CO} + V_{NO}I_{NO} = R_\Sigma I_\Sigma^2 = \frac{V_\Sigma^2}{R_\Sigma} = V_\Sigma I_\Sigma \tag{5.98}$$

It can be shown (see Appendix V) that

$$V_\Sigma^2 = \frac{1}{4}(V_{AN}^2 + V_{BN}^2 + V_{CN}^2 + V_{AB}^2 + V_{BC}^2 + V_{CA}^2) \tag{5.99}$$

thus if we rewrite (5.66) in a conventional way

$$S_\Sigma = 3V_e' I_e' = 3\frac{V_\Sigma}{\sqrt{3}}\frac{I_\Sigma}{\sqrt{3}} = V_\Sigma I_\Sigma \tag{5.100}$$

and the effective voltage and current are

$$V_e' = \sqrt{\frac{1}{12}(V_{AN}^2 + V_{BN}^2 + V_{CN}^2 + V_{AB}^2 + V_{BC}^2 + V_{CA}^2)} \tag{5.101}$$

$$I_e' = \sqrt{\frac{1}{3}(I_A^2 + I_B^2 + I_C^2 + I_N^2)} \tag{5.102}$$

Both V_e' and I_e' can be expressed in function of symmetrical components

$$V_e' = \sqrt{(V^+)^2 + (V^-)^2 + \frac{(V^0)^2}{4}} \tag{5.103}$$

and

$$I'_e = \sqrt{(I^+)^2 + (I^-)^2 + 4(I^0)^2} \tag{5.104}$$

A thorough mathematical demonstration [20, 24] for the maximum active power is found in Appendix VI where the effective voltage (7.36) is

$$V''_e = \sqrt{\frac{1}{9\rho + 3}[V_A^2 + V_B^2 + V_C^2 + \rho(V_{AB}^2 + V_{BC}^2 + V_{CA}^2)]}$$

$$= \sqrt{(V^+)^2 + (V^-)^2 + \frac{1}{3\rho + 1}(V^0)^2} \tag{5.105}$$

This expression is dependent on the ratio ρ and $V'_e = V''_e$ only for $\rho = 1.0$.

The method used in IEEE Std. 1459–2010 differs from the FBD-method mainly due to the fact that the compensated system, the one that transfers maximum active power, is perfectly symmetrical, void of any negative- or zero-sequence powers, Fig. 5.12d.

Assuming that the electrothermal effect is the deciding factor in determining the effective voltage V_e the following approach was proposed by the author [21]: the active power dissipated by the load under unity power factor conditions is assumed equal to the actual active power when the load is not compensated, (subunit power factor). If the metered consumer has some loads Δ-connected or floating neutral Y-connected, with a total power P_Δ and the remaining loads are Y-connected with four-wire system with a total power P_Y, then the total active power of the load can be expressed in the following manner

$$P = P_Y + P_\Delta = \frac{V_A^2 + V_B^2 + V_C^2}{R_Y} + \frac{V_{AB}^2 + V_{BC}^2 + V_{CA}^2}{R_\Delta} \tag{5.106}$$

where R_Y and R_Δ are equivalent resistances.

When the voltages V_A, V_B, and V_C are replaced with the symmetrical equivalent voltage source V_e, the active power supplied to the receiving end of the line must remain unchanged; in mathematical terms this means

$$P = P_Y + P_\Delta = \frac{3V_e^2}{R_Y} + \frac{9V_e^2}{R_\Delta} \tag{5.107}$$

Defining the ratio

$$\frac{P_\Delta}{P_Y} = \frac{9V_e^2}{R_\Delta} \frac{R_Y}{3V_e^2} = 3\frac{R_Y}{R_\Delta} = \xi \tag{5.108}$$

Both equations (5.106) and (5.107) give the total active power of the three-phase load. Substituting $R_\Delta = 3R_Y/\xi$ in (5.106) and (5.107) leads to the intermediary expression

$$\frac{V_A^2 + V_B^2 + V_C^2}{R_Y} + \frac{V_{AB}^2 + V_{BC}^2 + V_{CA}^2}{3R_Y}\xi = \frac{3V_e^2}{R_Y} + \frac{9V_e^2}{3R_Y}\xi = \frac{3V_e^2(1 + \xi)}{R_Y}$$

that helps to find the general expression of the effective voltage

$$V_e = \sqrt{\frac{3(V_A^2 + V_B^2 + V_C^2) + \xi(V_{AB}^2 + V_{BC}^2 + V_{CA}^2)}{9(1 + \xi)}} = \sqrt{(V^+)^2 + (V^-)^2 + \frac{(V^0)^2}{1 + \xi}} \quad (5.109)$$

Equation (5.109) (IEEE Std. 1459–2010) has to be discussed and compared with (5.101) (the FBD-method). Four cases of interest, each characterized by a different value of the ratio ξ, have been identified:

Case I: The four-wire system. No Δ-connected loads, no floating Y-connected loads, $P_\Delta = 0$, $R_\Delta \to \infty$ and $\xi = 0$.

$$V_e = \sqrt{\frac{V_A^2 + V_B^2 + V_C^2}{3}} = \sqrt{(V^+)^2 + (V^-)^2 + (V^0)^2} \quad (5.110)$$

Case II: The three-wire system. $P_Y = 0$, $R_Y \to \infty$, $\xi \to \infty$

$$V_e = \sqrt{\frac{(V_{AB}^2 + V_{BC}^2 + V_{CA}^2)}{9}} = \sqrt{(V^+)^2 + (V^-)^2} \quad (5.111)$$

This is equation (5.24). Expressions (5.110) and (5.111) define the highest and the lowest limits of V_e. Comparison with (5.101) and (5.103) shows that the zero-sequence voltage is for different equivalent load topologies.

Case III: Even load distribution, $P_\Delta = P_Y$, $R_\Delta = 3R_Y$ and $\xi = 1$.

$$V_e = \sqrt{\frac{3(V_A^2 + V_B^2 + V_C^2) + (V_{AB}^2 + V_{BC}^2 + V_{CA}^2)}{18}} = \sqrt{(V^+)^2 + (V^-)^2 + \frac{(V^0)^2}{2}} \quad (5.112)$$

Case IV: The particular case when $R_\Delta = R_Y$, $P_\Delta = 3P_Y$, $\xi = 3$.

$$V_e = V_e' = \sqrt{\frac{V_A^2 + V_B^2 + V_C^2 + V_{AB}^2 + V_{BC}^2 + V_{CA}^2}{12}} = \sqrt{(V^+)^2 + (V^-)^2 + \frac{(V^0)^2}{4}} \quad (5.113)$$

This expression is identical to (5.101) and (5.103).

In practical situations it is extremely difficult to determine the value of the ratio ξ. The loads are changing in time and the power network topology is quite involving. The initial issue of IEEE Std. 1459 assumed $\xi = 1.0$ and it is highly probable that this value will be kept unity for a long time. Fortunately, ξ affects only the contribution of the zero-sequence voltage V^0, and in practical power networks there are only traces of zero-sequence voltage. A large scale survey dedicated to power quality issues was conducted in the USA by EPRI and Electrotek

Concepts [22]. The data collected revealed that in medium and low voltage systems the ratio $100V^0/V^+$ will remain for 95% of the time smaller than 1.2%.

To estimate the difference between the values obtained from the FBD-method (5.101) and the IEEE Std. 1459–2010 (5.109), we may assume that the expression of V_e' (5.103) is the correct value and evaluate the normalized difference

$$\%\delta = 100\frac{V_e' - V_e}{V_e} \tag{5.114}$$

To simplify the computations the following notations are used:

$$\Upsilon^- = \frac{V^-}{V^+} \qquad \Upsilon^0 = \frac{V^0}{V^+}$$

From (5.103) and (5.109) we have

$$\frac{V_e'}{V^+} = \sqrt{1 + (\Upsilon^-)^2 + \frac{(\Upsilon^0)^2}{4}} \tag{5.115}$$

and

$$\frac{V_e}{V^+} = \sqrt{1 + (\Upsilon^-)^2 + \frac{(\Upsilon^0)^2}{1 + \xi}} \tag{5.116}$$

The largest difference occurs when $\Upsilon^- = 0$. In this case the percent difference is

$$\%\delta = 100\left[\sqrt{\frac{1 + \dfrac{(\Upsilon^0)^2}{1 + \xi}}{1 + \dfrac{(\Upsilon^0)^2}{4}}} - 1\right] \tag{5.117}$$

The results are plotted in Fig. 5.13. The solid curve corresponds to $V^0 = 0.04V^+$, a case that will not be encountered in power networks under normal operation and the dashed curve for $V^0 = 0.02V^+$, which is on the upper borderline of the actual zero-sequence voltage values. The maximum difference for $V^0/V^+ = 0.02$ is $\%\delta = 0.015\%$, i.e. 150 ppm. This value is found to be in the same range with some of the finest metering equipment errors.

In conclusion taking $\xi = 1$ or $\xi = 3$ will not cause major metering problems or discrepancies between the methods. The real difficulty is determining the value of ρ. This ratio can cause significant errors if it is ignored or incorrectly estimated [23]. More details on ρ estimation are given in Appendix VII.

A numerical example will help realize how close the two methods are. The circuit is presented in Fig. 5.14, the 60 Hz three-phase voltage supplying this system is not symmetrical and has the following symmetrical components:

$$\mathbf{V}_A^+ = 100\angle 0° \text{ V} \quad \mathbf{V}_A^- = 2.5\angle 0° \text{ V} \quad \mathbf{V}_A^0 = 1.2\angle -90° \text{ V}$$

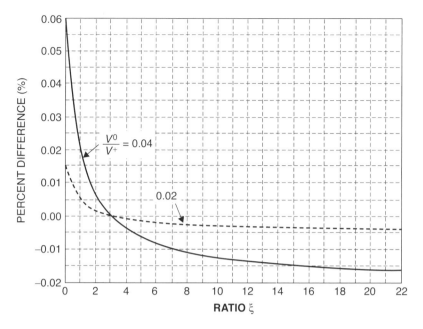

Figure 5.13 Percent difference between V_e and V'_e, (IEEE Std. 1459–2010, (1.109) and DIN 40110,(1.101)) in function of the ratio ξ.

The three line conductors have identical components, $R_s = 1.0\ \Omega$ and $L_s = 7.6$ mH. The neutral path has the resistance $R_N = 2.40\Omega$ and the inductance $L_N = 7.6$ mH. Part of the load is unbalanced, $R_A = 100\ \Omega$, $R_B = 500\ \Omega$ and $R_C = 1000\ \Omega$. The remaining load consists of three Δ-connected equal impedances, $\mathbf{Z}_L = 70 + j70.48\ \Omega$. On the load side the neutral conductor has a resistance $R_{NL} = 0.04\ \Omega$.

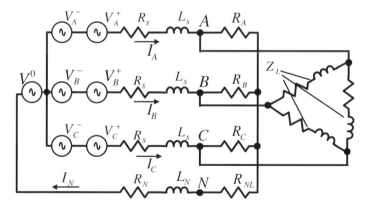

Figure 5.14 Example.

The currents and voltages measured at the load terminals are:

$$I_A = 3.561\angle - 38.283° \text{ A} \quad V_A = 91.354\angle - 5.231° \text{ V} \quad V_{AB} = 159.995\angle - 25.822° \text{ V}$$
$$I_B = 2.954\angle - 164.690° \text{ A} \quad V_B = 94.345\angle - 124.211° \text{ V} \quad V_{BC} = 155.222\angle - 92.376° \text{ V}$$
$$I_C = 2.819\angle - 74.763° \text{ A} \quad V_C = 90.066\angle - 121.164° \text{ V} \quad V_{CA} = 161.931\angle - 148.172° \text{ V}$$
$$I_N = 0.736\angle - 158.790° \text{ A}$$

The symmetrical components at the load terminals A, B, C, and N are

$$V^+ = 91.710 \text{ V} \qquad I^+ = 3.106 \text{ A}$$
$$V^- = 5.259 \text{ V} \qquad I^- = 0.274 \text{ A}$$
$$V^0 = 3.797 \text{ V} \qquad I^0 = 0.245 \text{ A}$$

To facilitate the comparison among different methods the negative- and the zero-sequence components are normalized in percent of the positive-sequence:

$$100 \frac{V^-}{V^+} = 4.140\% ; \qquad 100 \frac{V^0}{V^+} = 5.734\%$$

$$100 \frac{I^-}{I^+} = 8.812\% ; \qquad 100 \frac{I^0}{I^+} = 7.900\%$$

One will observe that in this example the zero-sequence voltage has a value that exceeds the level usually encountered in actual systems. The active power is

$$P = \Re e\{V_A I^*_A + V_B I^*_B + V_C I^*_C\} = 623.979 \text{ W}$$

I. The FBD-method leads to the following results:

$$I_\Sigma = \sqrt{I_A^2 + I_B^2 + I_C^2 + I_N^2} = 5.468 \text{ A}$$

$$V_\Sigma = \sqrt{\frac{1}{4}(V_A^2 + V_B^2 + V_C^2 + V_{AB}^2 + V_{BC}^2 + V_{CA}^2)} = 159.141 \text{ V}$$

and $V_e' = V_\Sigma/\sqrt{3} = 91.880$ V. The apparent power and the power factor are

$$S_\Sigma = V_\Sigma I_\Sigma = 870.133 \text{ VA} \qquad PF_\Sigma = \frac{P}{S_\Sigma} = 0.717$$

The branches' apparent powers are:

$$S_A = V_A I^*_A \quad S_B = V_B I^*_B \quad S_C = V_C I^*_C \quad S_N = V_N I^*_N$$

leading to the total active power

$$P = \Re e\{S_A + S_B + S_C + S_N\} = 623.979 \text{ W}$$

and reactive power

$$Q = \Im m\{\mathbf{S}_A + \mathbf{S}_B + \mathbf{S}_C + \mathbf{S}_N\} = 589.932 \text{ var}$$

The unbalance nonactive power is

$$F_{\Sigma u} = \sqrt{S_\Sigma^2 - P^2 - Q^2} = 140.572 \text{ var}$$

The balance conductance and susceptance are

$$G = \frac{P}{V_\Sigma^2} = 0.0246379 \text{ S} \qquad B = \frac{-Q}{V_\Sigma^2} = -0.0232935 \text{ S}$$

For the compensated system the hypothetical load plus compensator has an equivalent circuit that consists of four equal resistances $R_\Sigma = 1/G = 40.588 \ \Omega$ connected between the terminals A, B, C, and a common neutral point.

The new currents and voltages measured at the load terminals are:

$\mathbf{I}_A = 2.4592\angle - 4.109° \text{ A} \qquad \mathbf{V}_A = 99.799\angle - 4.109° \text{ V} \qquad \mathbf{V}_{AB} = 170.787\angle - 24.833° \text{ V}$

$\mathbf{I}_B = 2.376\angle - 125.108° \text{ A} \qquad \mathbf{V}_B = 96.420\angle - 125.108° \text{ V} \quad \mathbf{V}_{BC} = 164.424\angle - 93.942° \text{ V}$

$\mathbf{I}_C = 2.364\angle - 117.405° \text{ A} \quad \mathbf{V}_C = 95.921\angle - 117.405° \text{ V} \quad \mathbf{V}_{CA} = 170.787\angle - 147.283° \text{ V}$

$\mathbf{I}_N = 0.0216\angle - 86.059° \text{ A}$

The symmetrical components at the load terminals are

$$V^+ = 97.364 \text{ V} \qquad I^+ = 2.399 \text{ A}$$
$$V^- = 2.434 \text{ V} \qquad I^- = 0.021 \text{ A}$$
$$V^0 = 0.292 \text{ V} \qquad I^0 = 0.007 \text{ A}$$

Normalizing one finds

$$100 \frac{V^-}{V^+} = 2.50\% \ ; \qquad 100 \frac{V^0}{V^+} = 0.30\%$$
$$100 \frac{I^-}{I^+} = 2.50\% \ ; \qquad 100 \frac{I^0}{I^+} = 0.30\%$$

The apparent power equals the active power and the power factor becomes unity:

$$S_\Sigma = P = 701.240 \text{ W} \quad \text{thus} \quad PF = 1.0$$

II. The IEEE Std. 1459–2010 leads to the following results:

When we use $\rho = 1.0$ the effective current is $I_e = I_\Sigma/\sqrt{3} = 3.157$ A. The effective voltage is

$$V_e = \sqrt{\frac{3(V_A^2 + V_B^2 + V_C^2) + V_{AB}^2 + V_{BC}^2 + V_{CA}^2}{18}} = 91.890 \text{ V}$$

yielding the effective apparent power

$$S_e = 3V_e I_e = 870.228 \text{ VA}$$

and the power factor

$$PF_e = \frac{P}{S_e} = 0.717$$

The active power is $P = 623.979$ W. The positive-sequence apparent power is

$$S^+ = 3V^+ I^+ = 854.646 \text{ VA}$$

From (5.89) and (5.92) results

$$S_u = \sqrt{S_e^2 - S_u^2} = 166.375 \text{ VA}$$

and

$$N_u = \sqrt{S_e^2 - P^2 - Q^2} = 145.438 \text{ var} \approx F_{\Sigma u} = 140.572 \text{ var}$$

In the IEEE approach the load is compensated in such a manner that the line currents are sinusoidal, with equal amplitudes and 120° out of phase. However, since the supplied voltages are not symmetrical, the compensated line currents can not be exactly in-phase with the line-to-neutral voltages. A sound approach is to design a compensator that brings the line currents in-phase with the respective positive-sequence voltage. The line current amplitude has to be adjusted to a level that ensures minimum active power flow into the compensator.

If ρ ratio is correctly considered, $\rho = R_N/R_s = 2.40/1.0 = 2.40$, the effective current value increases to $I_e = 3.197$ A, causing the apparent power to increase 1.26% over the apparent power that corresponds to $\rho = 1.0$. Consequently the power factor decreases to 0.707. For the system compensated according to IEEE the currents and voltages measured at the load terminals are:

$\mathbf{I}_A = 2.257\angle -4.499°$ A $\quad \mathbf{V}_A = 100.020\angle -4.280°$ V $\quad \mathbf{V}_{AB} = 171.218\angle 25.105°$ V

$\mathbf{I}_B = 2.257\angle -124.500°$ A $\quad \mathbf{V}_B = 97.347\angle -124.610°$ V $\quad \mathbf{V}_{BC} = 164.458\angle -93.786°$ V

$\mathbf{I}_C = 2.257\angle 115.50°$ A $\quad \mathbf{V}_C = 95.018\angle 117.880°$ V $\quad \mathbf{V}_{CA} = 170.742\angle 147.613°$ V

$\mathbf{I}_N = 0$ A

$$(5.118)$$

The symmetrical components at the load terminals are

$$V^+ = 97.250 \text{ V} \qquad I^+ = 2.257 \text{ A}$$
$$V^- = 5.568 \text{ V} \qquad I^- = 0 \text{ A}$$
$$V^0 = 3.778 \text{ V} \qquad I^0 = 0 \text{ A}$$

The currents are practically pure positive-sequence; the voltages, however, remain asymmetrical with the normalized values

$$100 \frac{V^-}{V^+} = 5.725\% \; ; \qquad 100 \frac{V^0}{V^+} = 3.88\%$$

Since in this case the zero-sequence current is nil, the value of ρ is inconsequential. The effective voltage and current are

$$V_e = 97.483 \text{ V} \qquad I_e = 2.257 \text{ A}$$

yielding

$$S_e = 660.093 \text{ VA} \; ; \qquad P = 657.140 \text{ W} \quad \text{and} \quad PF = 0.995$$

Table 5.1 enables the comparison of the methods:

Table 5.1 Comparison among different methods

	VECTOR	DIN	IEEE ($\rho = 1.0$)	IEEE ($\rho = 2.40$)
V_e, V_e' (V)	—	91.8803	91.8904	91.8844
I_e (A)	—	3.1567	3.1567	3.1966
S_V, S_Σ, S_e (VA)	857.194	870.1326	870.228	881.144
PF	0.727	0.716	0.716	0.706
$\frac{I^-/I^+}{Compensated}$ (%)	—	2.50	0	0
$\frac{I^0/I^+}{Compensated}$ (%)	—	0.30	0	0
$\frac{V^-/V^+}{Compensated}$ (%)	—	2.50	5.725	5.725
$\frac{V^0/V^+}{Compensated}$ (%)	—	0.30	3.88	3.88

It is learned from this example that if it is assumed that $\rho = 1.0$ both methods, DIN and IEEE, give practically the same results as concerns I_e, V_e, S_e, and the PF. The post-compensation results are different: The IEEE approach yields a perfect sinusoidal, positive-sequence currents; the DIN currents are tainted by residual symmetrical components. The IEEE method does not help reduce significantly the zero- and negative-sequence voltages (V^-/V^+ from 5.73% to 5.72% and V^0/V^+ from 5.73% to 3.88%). The DIN seems more effective causing for V^0/V^+

a reduction from 5.73% to 0.3% and for V^-/V^+ from 4.14% to 2.50%. High negative-sequence voltage is damaging motors, but high zero-sequence currents cause a host of unwanted problems (stray voltage, electromagnetic interference, relay malfunction, etc.).

If the supplier of energy is involved in the process of system balancing and imposes on all the large and medium consumers to improve the loads balance and to aim for unity power factor, then both methods converge toward the same excellent symmetry of currents and voltages and the disappearance of the residual currents.

To further emphasize the difference between the methods, the function

$$\Delta P = R_s(I_A^2 + I_B^2 + I_C^2 + \rho I_N^2) = F(S^2)$$

is observed for the three-phase circuit shown in Fig. 5.15a. The load resistance R varies in the range $1 \leq R \leq 1000$ Ω. The normalized apparent powers squared, $(S_V/S_0)^2$, $(S_\Sigma/S_0)^2$, $(S_e/S_0)^2$ for $\rho = 1$ and $(S_e/S_0)^2$ for the correct $\rho = 3$ are plotted in function of the normalized power loss $\Delta P/\Delta P_0$, where $S_0 = S_e$ for $R = 40$ Ω.

The results shown in Fig. 15b prove the total lack of linearity of the vector apparent power and the reasonable linearity of S_e^2 for $\rho = 3$ followed by S_e^2 with $\rho = 1$ and the S_Σ^2.

An alert reader will ask the following question: "What happens with (5.109) when we deal with a consumer with severely unbalanced loads, such as a V-connected transformer or a single line-to-neutral connected load?"

Evidently the consumer is supplied with a three-phase, three-wire system. The V-connected load has the active power P_V. The next step is to replace the V-connection with a Δ-connection that consists of three resistances R_Δ. The equivalence of active power between the V and the Δ connection gives

$$P_V = \frac{V_{AB}^2 + V_{BC}^2 + V_{CA}^2}{R_\Delta}$$

Finally, the equivalence with the hypothetical symmetrical three-phase line supplied with the line-to-neutral voltage V_e leads to

$$P_V = 9V_e^2/R_\Delta$$

hence

$$V_e = \sqrt{\frac{R_\Delta P_V}{9}} = \sqrt{\frac{R_\Delta(V_{AB}^2 + V_{BC}^2 + V_{CA}^2)}{9P_V R_\Delta}P_V} = \sqrt{\frac{V_{AB}^2 + V_{BC}^2 + V_{CA}^2}{9}}$$

an expression identical to (5.111).

In a similar way one may approach the case of a single-phase load connected line-to-neutral. In this case the active power is

$$P_R = \frac{V_A^2}{R_R} = \frac{V_A^2 + V_B^2 + V_C^2}{R_Y} = \frac{3V_e^2}{R_Y}$$

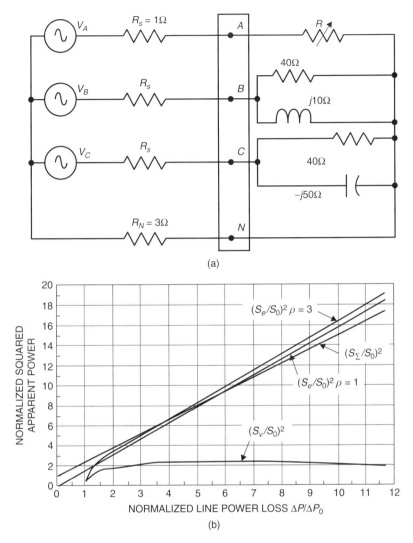

Figure 5.15 Comparison among different definitions: (a) Circuit used to compute the apparent powers measured at the terminals A, B, C, N. (b) $(S_V/S_0)^2$, $(S_\Sigma/S_0)^2$, $(S_e/S_0)^2$ for $\rho = 1$ and $(S_V/S_0)^2$ for $\rho = 3$ vs. $\Delta P/\Delta P_0$.

where R_R is the actual equivalent resistance, connected A to neutral and R_Y is one of the three equal resistances that make the equivalent Y-connected load supplied with symmetrical V_e, line-to-neutral voltages.

From the last equations we find

$$V_e = \sqrt{\frac{V_A^2 + V_B^2 + V_C^2}{3}}$$

In the case of more involving, unbalanced three-phase loads, the expressions for V_e, (5.110) and (5.111), give satisfactory results. For three-wire systems use (5.110) and for four-wire systems use (5.111). The expression (5.109) can be used if the load topology is well known and the voltages and the active powers supplied to the loads are correctly measured. The main issue is the load separation in Δ- and Y-connected equivalent loads, that dissipate the same active power as the original load, i.e.

$$R_\Delta = \frac{V_{AB}^2 + V_{BC}^2 + V_{CA}^2}{P_\Delta} \quad \text{and} \quad R_Y = \frac{V_A^2 + V_B^2 + V_C^2}{P_Y}$$

The following example helps to clarify such a situation: A three-phase load consists of three equal resistances $R = 20\Omega$, Δ-connected and one resistance R_R, connected phase A-to-neutral, that takes two values, $R_R = 10\Omega$ and $R_R = 1\Omega$. The measured rms voltages are: Line-to-neutral: $\mathbf{V}_A = 350.00\angle 0°$ V, $\mathbf{V}_B = 365.00\angle -116°$ V, $\mathbf{V}_C = 372.00\angle 124°$ V, Line-to-line: $V_{AB} = 606.30$ V, $V_{BC} = 638.26$ V, $V_{CA} = 637.57$ V.

For $R_R = 10\Omega$, computations yield: $R_\Delta = 20\Omega$, $R_Y = 32.17\Omega$ and $\xi = 3R_Y/R_\Delta = 4.826$. For $R_R = 1\Omega$ results $R_Y = 3.217\Omega$ and $\xi = 0.483$. The values of V_e for expression (5.109) and the four cases presented in section 5.54, see (5.110), (5.111), (5.112) and (5.113) are listed in the following table:

Table 5.2 Computed Values of V_e Using the Expressions Recommended by IEEE Std. 1459–2010 and DIN 40110

Equation number	(5.109)	(5.110)	(5.111)	(5.112)	(5.113)
$R_R = 10\ \Omega$	362.341	362.450	362.318	362.384	362.384
$R_R = 1\ \Omega$	362.407	362.450	362.450	362.384	362.351
	IEEE	Case I	Case II	Case III	DIN

The results are very close, the difference between the extremes being less than 0.04%.

5.6 Problems

5.1 The active and reactive powers of a three-phase unbalanced load are as follows:

$$P_a = P_b = 2.0\ \text{kW} \qquad Q_a = Q_b = 1.0\ \text{kvar}$$
$$P_c = 3.0\ \text{kW} \qquad Q_c = -2.0\ \text{kvar}$$

Compute the arithmetic and vector apparent powers and the respective power factors.

5.2 Assuming that the four conductors that supply the load from the previous problem have equal resistances, compute the Buchholz-Goodhue effective apparent power $S_e = 3V_e I_e$ and the power factor. The rms line-to-neutral voltages are $V_a = V_b = 120$ V and $V_c = 110$ V.

5.3 A three-phase load consists of three Y-connected resistances, R, R, and ηR supplied by a four-wire system. At the load terminals the line-to-neutral voltages are $V\angle 0°$, $V\angle -120°$ and $V\angle 120°$. The coefficient η controls the load unbalance. Assuming that all four conductors have equal resistances R_s, plot the graphs describing the normalized power loss $\Delta P/\Delta P_0$ versus $(S_e/S_0)^2$, $(S_V/S_0)^2$ and $(S_A/S_0)^2$ when $0.1 < \eta < 50$. The base values ΔP_0 and S_0 are the values obtained when $\eta = 1.0$.

5.4 Repeat problem 5.3 for the case when the neutral conductor has a resistance $R_n = 2.3R_s$. Explain your results.

5.5 Following are the voltage and current phasors measured at the terminals of a three-phase load:

$$\mathbf{V}_a = 650\angle 0° \qquad \mathbf{V}_b = 670\angle -128° \quad \mathbf{V}_c = 659\angle 127°$$
$$\mathbf{I}_a = 150\angle -25° \qquad \mathbf{I}_b = 100\angle -151° \quad \mathbf{I}_c = 659\angle 92°$$

Determine the power triangles $P_k \ Q_k \ S_k \quad k = a, \ b, \ c$ and compute the arithmetic and the vector apparent powers. Find the neutral current I_n, the effective current, and voltage I_e and V_e. Compute the apparent power S_e. Assume $R_n = R_s$, i.e. $\rho = 1.0$.

Compute the symmetrical components of the voltage and current. Prove numerically that $P = P^+ + P^- + P^0$, $Q = Q^+ + Q^- + Q^0$ and $S_V = |\mathbf{S}^+ + \mathbf{S}^- + \mathbf{S}^0|$.

5.6 Use Lagrange multipliers to find the maximum active power that can be transmitted to a three-phase load via a three-wire system. The line-to-line voltage phasors are given and so are the three line rms currents.

5.7 An ideal three-phase three-wire cable supplies a load that consists of two identical heaters (each having the resistance R) connected line-to-line (A to B and B to C). The load voltages ($\sqrt{3}V$ rms line-to-line) are perfectly symmetrical. Sketch the phasors diagram, compute I_e, P, S_e, N, and the PF. Compare with the S_V and PF_V. Assuming the cable coaxial, with the radii a, b, and c, determine the actual flow of power from source to load. Assuming that this cable has very small, but equal resistances $R_s \ll R$, on each line. Check if the line power loss $\Delta P = S^2/(3R_s V^2)$ for S_e and for S_V.

5.8 An unbalanced load is supplied by a three-phase four-wire cable. The neutral wire impedance is negligible. The three lines have equal resistances r. The line reactances can be also neglected. The line currents are:

$$I_A = \beta I \angle -\theta ; \qquad I_B = I \angle -120° -\theta ; \qquad I_C = I \angle -120° -\theta$$

with the coefficient $\beta \neq 1.0$. Show that $P^+ - P_{out} = -(P^- + P^0)$. Where P^+, P^-, P^0 are the symmetrical components active powers measured at the load's terminals and P_{out} is the power delivered by the load, in this case converted in heat.

5.9 Compute V_e and I_e for a three-phase supplied with the voltages $V_A = 100$ V, $V_B = 80$ V, $V_C = 80$ V, $V_{AB} = 160.64$ V, $V_{BC} = 129.44$ V and $V_{CA} = 160.64$ V. The load consists of heating elements; three equal resistances $R_\Delta = 9\Omega$ connected in Δ and one resistance $R_R = 5\Omega$ connected line-to-neutral, phase A to neutral. Hint: Assume equivalent Y-connected resistances that represent R_R, i.e.

$$(V_A^2 + V_B^2 + V_C^2)/R_Y = V_A^2/R_R$$

5.7 References

[1] "Power Factor in Polyphase Circuits," *Trans AIEE* Vol. 39, July 1, 1920, pp. 1449–540.

[2] Lincoln P. M.: "Polyphase Power Factor," *Trans AIEE*, Vol. 39, 1920, pp. 1477–79.

[3] Fortescue C. L.: "Polyphase Power Representation by Means of Symmetrical Coordinates," *Trans AIEE*, Vol. 39, 1920, pp. 1481–89.

[4] Czarnecki L. S.: "Energy Flow and Power Phenomena in Electrical Circuits: Illusions and Reality," *Archiv für Elektrotechnik*, Vol. 82, 2000, pp. 119–26.

[5] Buchholz F.: "Die Drehstrom-Sheinleistung bei Ungleichmassiger Belastung Der Drei Zweige," *Licht und Kraft* No. 2, Jan. 1922, pp. 9–11.

[6] Knowlton A. E.: "Reactive Power in Need of Clarification," *Trans AIEE*, Vol. 52, Sept. 1933, pp. 744–805.

[7] Goodhue W. M.: Discussion to [6]: *Trans AIEE*, Vol. 52, Sept. 1933, p. 787.

[8] Emanuel A. E.: "The Buchholz-Goodhue Apparent Power Definition: The Practical Approach for Nonsinusoidal and Unbalanced Systems," *IEEE Transactions on Power Delivery*, Vol. 13, No. 2, April 1998, pp. 344–50.

[9] IEEE Std 1459–2000, "*IEEE Trial-Use Standard Definitions for the Measurement of Electric Power Quantities Under Sinusoidal, Nonsinusoidal, Balanced or Unbalanced Conditions*." (Upgraded to Full-Use, August 2002.)

[10] Emanuel A. E.: "On the Definition of Power Factor and Apparent Power in Unbalanced Polyphase Circuits with Sinusoidal Voltages and Currents," *IEEE Transactions on Power Delivery*, Vol. 8, No. 3, July 1993, pp. 841–47.

[11] Lurye L. S.: "The Computation of Powers in Three-Phase Systems," *Elektritchestvo* Vol. 1, 1951, pp. 47–52.

[12] Manea F.: "The Powers of Unbalanced and Nonsinusoidal Three-Phase Systems," *Studies and Research of Energetics* (Etudea et Recherches D'Energetique) Vol. 4, 1960, pp. 771–77.

[13] Depenbrock M.: "*Untersuchungen uber die Spanungs- und Leistungsverhaltnisse bei Umrichtern ohne Energiespeicher,*" Disertation, TH Hannover, 1962.

[14] Depenbrock M.: "The FBD-Method, a Generally Applicable Tool for Analysing Power Relations," *IEEE Transactions on Power Systems*, Vol. 8, No. 2, 1993, pp. 381–86.

[15] Buchholz F.: "*Das Begriffsystem Rechtleistung, Wirkleistung, Totale Blindleistung,*" Selbstverlag, Munchen, 1950.

[16] DIN 40110 AC Quantities, Part 1: Single-Phase Circuits, March 1994. Part 2: Polyphase Circuits, November 2002 (in German).

[17] Czarnecki L. S.: "Orthogonal Decomposition of the Currents in a 3-Phase Non-Linear Asymmetrical Circuit with a Nonsinusoidal Voltage Source," *IEEE Transactions on Instrumentation and Measurements*, Vol. 37, No. 3, 1988, pp. 30–34.

[18] Depenbrock M.: "Quantities of a Multiterminal Circuit Determined on the Basis of Kirchhoff's Laws," *ETEP*, Vol. 8, 1998, pp. 249–57.

[19] Ferrero A.: "Definitions of Electrical Quantities Commonly Used in Nonsinusoidal Conditions," *ETEP*," Vol. 8, 1998, pp. 235–40.

[20] Willems J. L., Ghijselen J. A., Emanuel A. E.: "The Apparent Power Concept and the IEEE Standard 1459–2000," *IEEE Transactions on Power Delivery*, Vol. 20, No. 2, April 2005, pp. 876–84.

[21] Emanuel A. E.: "Summary of IEEE Standard 1459: Definitions for the Measurement of Electric Quantities Under Sinusoidal, Nonsinusoidal, Balanced or Unbalanced Conditions," *IEEE transactions on Industry Applications*, Vol. 40, No. 3, May/June 2004, pp. 869–76.

[22] EPRI-RP3098-01, "*An Assessment of Distribution System Power Quality.*" Prepared by Electrotek Concepts, February 1996.

[23] Pajic S., Emanuel A. E.: "Effect of Neutral Path Power Losses on the Apparent Power Definitions: A Preliminary Study," *IEEE transactions on Power Delivery*, Vol. 24, No. 2, April. 2009, pp. 517–23.

[24] Pajic S., Emanuel A. E.: "Modern Apparent Power Definitions: Theoretical Versus Practical Approach—The General Case," IEEE Transactions on Power Delivery, Vol. 21, No. 4, Oct. 2006, pp. 1787–92.

[25] Spath H.: "A General Purpose Definition of Active Current and Non-Active Power Based on the German Standard DIN 40110," Electrical Engineering, No. 89, 2007, pp. 167–75.

6

Three-Phase Nonsinusoidal and Unbalanced Conditions

your old men will dream dreams,
your young men will see visions.

—Joel 2:28

This chapter deals with the most complex case, poly-phase circuits with distorted waveforms, unbalanced loads, and asymmetrical voltages. The literature abounds with studies of such conditions [1–12] presenting approaches that complete or contradict each other. Unfortunately no universal agreement has been achieved. The theoreticians are way ahead of the instrumentation manufacturers [13] who, for economical reasons, continue to uphold the Silsbee and Curtis approach [14]. The definitions of powers officially adopted for the first time in 1941 [15] were promoted for a long time in the IEEE Standard 100 [16]. This chapter details only the most popular definitions, either in use by the electric utilities, or recommended by major standards.

6.1 The Vector Apparent Power Approach

In 1935 Silsbee and Curtis [14] expanded Budeanu's single-phase approach to three-phase systems. Their method can be easily understood from the three-dimensional representation shown in Fig. 6.1 The vector apparent power S_V is the diagonal of the box with the sides P, Q_B, and D_B, i.e.

$$S_V = \sqrt{P^2 + Q_B^2 + D_B^2} \tag{6.1}$$

where

$$P = P_A + P_B + P_C = \sum_{k=A,B,C} P_k \tag{6.2}$$

Power Definitions and the Physical Mechanism of Power Flow   Alexander Eigeles Emanuel
© 2010 John Wiley & Sons, Ltd

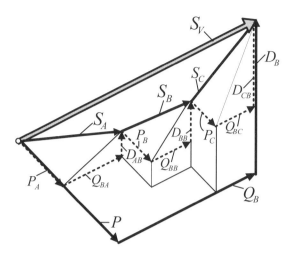

Figure 6.1   Vector apparent power resolution.

is the total three-phase's load active power, and

$$P_k = \sum_h V_{kh} I_{kh} \cos(\theta_h) \tag{6.3}$$

is the total active power of phase k, θ_h is the phase shift between the harmonic current and harmonic voltage phasors of order h

$$Q_B = Q_{BA} + Q_{BB} + Q_{BC} = \sum_{k=A,B,C} Q_{Bk} \tag{6.4}$$

is the total three-phase's load reactive power according to Budeanu, and

$$Q_{Bk} = \sum_h V_{kh} I_{kh} \sin(\theta_h) \tag{6.5}$$

is the total Budeanu's reactive power of phase k.

The distortion power of phase k is computed from the expression

$$D_k = \sqrt{S_k^2 - P_k^2 - Q_k^2} \tag{6.6}$$

where

$$S_k = V_k I_k = \sqrt{\sum_h V_{kh}^2 \sum_h I_{kh}^2} \tag{6.7}$$

is the apparent power of phase k.

Instruments based on the vector apparent power were implemented in the last two decades. Such meters are based on Budeanu reactive power, hence using a definition that is rejected by many modern engineers. Moreover, an incorrect value of Q_B also leads to an incorrect value of D_B.

6.2   The IEEE Std. 1459–2010's Approach

The actual three-phase supplying line carries the current harmonics I_{ah}, I_{bh}, and I_{ch}. In the case of a four-wire system, there is also a residual or neutral current harmonic I_{nh}. The actual line and load are replaced with a hypothetical, perfectly compensated system that draws perfectly sinusoidal positive-sequence currents, I_e and a nil neutral current. The line power loss in the hypothetical system equals the actual power loss causing the same thermal stress. Mathematically this translates in the equality

$$\Delta P = 3r_e I_e^2 = r_{dc} \sum_h K_{sh} (I_{ah}^2 + I_{bh}^2 + I_{ch}^2) + r_{ndc} \sum_h K_{snh} I_{nh}^2 \qquad (6.8)$$

where

K_{sh} and K_{snh} are the h-harmonic order combined skin and proximity effect coefficients for the line conductors and the neutral current path, respectively,

r_{dc} and r_{ndc} are the line and neutral current path dc resistances,

$r_e = K_{s1} r_{dc}$ is the equivalent resistance labeled R_s in the previous chapters. K_{s1} is the combined skin and proximity effect at the power frequency (60 or 50 Hz).

The effective current is obtained from (6.8):

$$I_e = \sqrt{\frac{1}{3}\left\{\sum_h \left[\frac{K_{sh}}{K_{s1}} (I_{ah}^2 + I_{bh}^2 + I_{ch}^2) + \frac{K_{snh}}{K_{s1}}\frac{r_{ndc}}{r_{dc}} I_{nh}^2\right]\right\}} \qquad (6.9)$$

The effective current has two orthogonal components:

$$I_e = \sqrt{I_{e1}^2 + I_{eH}^2} \qquad (6.10)$$

where

$$I_{e1} = \sqrt{\frac{1}{3}\left[(I_{a1}^2 + I_{b1}^2 + I_{c1}^2) + \rho_1 I_{n1}^2\right]} \; ; \quad \rho_1 = \frac{K_{sn1}}{K_{s1}}\frac{r_{ndc}}{r_{dc}} \qquad (6.11)$$

is the fundamental currents contribution and

$$I_{eH} = \sqrt{\frac{1}{3}\left\{\sum_{h\neq1} \left[K_h (I_{ah}^2 + I_{bh}^2 + I_{ch}^2) + \rho_h I_{nh}^2\right]\right\}} \; ; \quad K_h = \frac{K_{sh}}{K_{s1}} \quad \rho_h = \frac{K_{snh}}{K_{s1}}\frac{r_{ndc}}{r_{dc}} \qquad (6.12)$$

is the current harmonics contribution.

In practice the ratios ρ_1, ρ_h and K_h are not well known. Network topology changes, temperature changes, and seasonal changes in soil humidity and temperature make the estimation of these ratios a very difficult task. The IEEE Std. 1459–2010 recommends using the values

$\rho_1 = \rho_h = K_h = 1.0$. This simplification does not penalize the consumer and leads to the following expressions:

$$I_e = \sqrt{\frac{1}{3}(I_a^2 + I_b^2 + I_c^2 + I_n^2)} \tag{6.13}$$

$$I_{e1} = \sqrt{\frac{1}{3}(I_{a1}^2 + I_{b1}^2 + I_{c1}^2 + I_{n1}^2)} \tag{6.14}$$

$$I_{eH} = \sqrt{\frac{1}{3}(I_{aH}^2 + I_{bH}^2 + I_{cH}^2 + I_{nH}^2)} = \sqrt{I_e^2 - I_{e1}^2} \tag{6.15}$$

where

$$I_{\kappa H}^2 = \sum_{h \neq 1} I_{\kappa h}^2 \; ; \quad \kappa = a, b, c, n \tag{6.16}$$

For a three-wire system $I_{n1} = I_{nh} = 0$ and the above expressions become

$$I_e = \sqrt{\frac{1}{3}(I_a^2 + I_b^2 + I_c^2)} \tag{6.17}$$

$$I_{e1} = \sqrt{\frac{1}{3}(I_{a1}^2 + I_{b1}^2 + I_{c1}^2)} \tag{6.18}$$

$$I_{eH} = \sqrt{\frac{1}{3}(I_{aH}^2 + I_{bH}^2 + I_{cH}^2)} = \sqrt{I_e^2 - I_{e1}^2} \tag{6.19}$$

A similar procedure is used to define the equivalent voltage V_e: the compensated hypothetical load has a unity, or close to unity, power factor. This means that only active power is supplied to the line end. The load is separated in Δ-connected loads that are supplied with the active power P_Δ (this includes also the floating neutral Y-connected loads) and the Y-connected loads with the active power P_Y (this includes all the loads connected to neutral). The Δ-connected loads are balanced and characterized by equivalent line-to-line resistances R_Δ. Similarly the Y-connected loads are represented by means of a balanced load with three line-to-neutral resistances R_Y. The equivalence of active power between the actual and the hypothetical system is

$$\frac{3V_e^2}{R_Y} + \frac{9V_e^2}{R_\Delta} = \frac{\sum_h (V_{ah}^2 + V_{bh}^2 + V_{ch}^2)}{R_Y} + \frac{\sum_h (V_{abh}^2 + V_{bch}^2 + V_{cah}^2)}{R_\Delta} \tag{6.20}$$

The notation

$$\xi = \frac{P_\Delta}{P_Y} = \frac{9V_e^2}{R_\Delta} \frac{R_Y}{3V_e^2} = \frac{3R_Y}{R_\Delta}$$

helps rewrite (6.20) as follows:

$$\frac{3(1+\xi)}{R_Y}V_e^2 = \frac{1}{R_Y}\left\{\frac{\sum_h\left(V_{ah}^2+V_{bh}^2+V_{ch}^2\right)}{1} + \frac{\sum_h\left(V_{abh}^2+V_{bch}^2+V_{cah}^2\right)}{3/\xi}\right\} \quad (6.21)$$

From here we find the effective voltage

$$V_e = \sqrt{\frac{3\sum_h\left(V_{ah}^2+V_{bh}^2+V_{ch}^2\right)+\xi\sum_h\left(V_{abh}^2+V_{bch}^2+V_{cah}^2\right)}{9(1+\xi)}} \quad (6.22)$$

The separation of fundamental components from the harmonics and interharmonics using $V_e^2 = V_{e1}^2 + V_{eH}^2$ leads to *the fundamental effective voltage*

$$V_{e1} = \sqrt{\frac{3\left(V_{a1}^2+V_{b1}^2+V_{c1}^2\right)+\xi\left(V_{ab1}^2+V_{bc1}^2+V_{ca1}^2\right)}{9(1+\xi)}} = \sqrt{\left(V_1^+\right)^2+\left(V_1^-\right)^2+\frac{\left(V_1^0\right)^2}{1+\xi}} \quad (6.23)$$

an expression identical to (5.109) for which the IEEE Std. 1459–2010 recommends $\xi = 1$.
The second term is the *harmonic effective voltage*

$$V_{eH} = \sqrt{\frac{3\sum_{h\neq1}\left(V_{ah}^2+V_{bh}^2+V_{ch}^2\right)+\xi_h\sum_{h\neq1}\left(V_{abh}^2+V_{bch}^2+V_{cah}^2\right)}{9(1+\xi_h)}}$$

$$= \sqrt{\sum_{h\neq1}\left[\left(V_h^+\right)^2+\left(V_h^-\right)^2+\frac{\left(V_h^0\right)^2}{1+\xi_h}\right]} = \sqrt{V_e^2-V_{e1}^2} \quad (6.24)$$

Again we are faced with the difficulty of ξ_h determination. Typically for $h = 3n \pm 1$, $n = 0, 1, 2, 3, \ldots$, $V_h^0 \ll V_h^+$ and $\xi = 1$ is a satisfactory compromise. However, for $h = 3n$, $V_h^0 \gg V_h^+$ and an error may take place if one assumes $\xi = 1$.

The theoretically correct approach is to define for each harmonic an effective current and voltage, I_{eh} and V_{eh}, with a ρ_h and a ξ_h. Fortunately (6.24) offers a shortcut, $V_{eH} = \sqrt{V_e^2-V_{e1}^2}$.

The resolution of $S_e = 3V_eI_e$ is done in the same way as it was in section 4.5: the effective power is separated into two major terms

$$S_e^2 = S_{e1}^2 + S_{eN}^2$$

where

$$S_{e1} = 3V_{e1}I_{e1}$$

is the *fundamental, or 60/50 Hz, effective apparent power* and the term S_{eN} is the *nonfundamental effective apparent power.* In turn S_{eN} has three components

$$S_{eN}^2 = S_e^2 - S_{e1}^2 = D_{eI}^2 + D_{eV}^2 + S_{eH}^2 \qquad (6.25)$$

where

$$D_{eI} = 3V_{e1}I_{eH} \qquad (6.26)$$

is the *current distortion power*, usually the largest component of S_{eN},

$$D_{eV} = 3V_{eH}I_{e1} \qquad (6.27)$$

is the *voltage distortion power* and

$$S_{eH} = 3V_{eH}I_{eH} \qquad (6.28)$$

is the *effective harmonic apparent power*. Two components characterize S_{eH},

$$S_{eH}^2 = P_H^2 + D_{eH}^2 \qquad (6.29)$$

Here

$$P_H = \sum_{h \neq 1} \{V_{ah}I_{ah}\cos(\theta_{ah}) + V_{bh}I_{bh}\cos(\theta_{bh}) + V_{ch}I_{ch}\cos(\theta_{ch})\} \qquad (6.30)$$

is the total *harmonic active power* and

$$D_{eH} = \sqrt{S_{eH}^2 - P_H^2} \qquad (6.31)$$

is the *harmonic distortion power*.

The components of S_{eN} can be expressed in function of the *equivalent total harmonic distortions*

$$THD_{eV} = \frac{V_{eH}}{V_{e1}} \text{ for voltage and } THD_{eI} = \frac{I_{eH}}{I_{e1}} \text{ for current} \qquad (6.32)$$

From (6.25) results

$$S_{eN}^2 = \left[\frac{D_{eI}^2}{S_{e1}^2} + \frac{D_{eV}^2}{S_{e1}^2} + \frac{S_{eH}^2}{S_{e1}^2}\right]S_{e1}^2 = \left[\frac{(V_{e1}I_{eH})^2}{(Ve1I_{e1})^2} + \frac{(V_{eH}I_{e1})^2}{(Ve1I_{e1})^2} + \frac{(V_{eH}I_{eH})^2}{(Ve1I_{e1})^2}\right]S_{e1}^2 \qquad (6.33)$$

and substitution of (6.32) in (6.33) gives a practical expression

$$S_{eN} = S_{e1}\sqrt{THD_{eI}^2 + THD_{eV}^2 + THD_{eI}^2\,THD_{eV}^2} \qquad (6.34)$$

which helps to evaluate separately the contributions of the three terms of S_{eN} to the harmonic pollution; $D_{eI} = (THD_{eI})S_{e1}$, $D_{eV} = (THD_{eV})S_{e1}$ and $S_{eH} = (THD_{eI})(THD_{eV})S_{e1}$.

The most important powers are the *fundamental (60/50 Hz) positive-sequence active and reactive powers* P_1^+ and Q_1^+, respectively. They are tied to the *fundamental positive-sequence apparent power*

$$\left(S_1^+\right)^2 = \left(P_1^+\right)^2 + \left(Q_1^+\right)^2$$

with the *fundamental or the 60/50 Hz positive-sequence power factor*

$$PF_1^+ = \frac{P_1^+}{S_1^+}$$

The *power factor* follows its classical definition

$$PF = \frac{P}{S_e} = \frac{P_1 + P_H}{S_e}$$

In some situations there may be interest in evaluating symmetrical components such as fundamental zero- and negative-sequence currents, voltages, and powers. The same may apply for the harmonic current and voltage imbalance and separation by symmetrical components. The components and the subcomponents discussed in this section are summarized in Fig. 6.2. The unbalance fundamental power

$$S_{U1} = \sqrt{S_{e1}^2 - \left(S_1^+\right)^2}$$

is introduced to allow the positive-sequence powers P_1^+ and Q_1^+ separation from S_{e1}. S_{U1} includes the contributions of the 50/60 Hz negative- and positive–sequence powers and gives a crude indication about the degree of load imbalance.

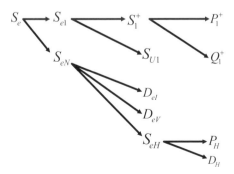

Figure 6.2 The powers' tree (IEEE Std. 1459–2010).

6.3   The DIN 40110's Approach

The definitions used in this standard are based on a simple concept: the compensated load plus compensator of a four-wire system are electrically equivalent to four linear and equal resistances R_Σ connected to a common virtual point O (see Fig. 5.12c). A three-wire system needs three resistances. Evidently such a virtual load has a unity power factor under any circumstances, but when the voltage is distorted or asymmetrical so will the currents. Since the voltages are usually only lightly unbalanced and distorted, the results of such compensations are satisfactory.

The procedure described in the German Standard DIN 40110 is based on the FBD-method [18]. It follows the same ideas presented in sections 5.5.1 and 5.5.4. The standard pivots around the expression

$$S_\Sigma^2 = P_\Sigma^2 + Q_{tot\Sigma}^2 \tag{6.35}$$

where the subscript Σ denotes collective value; S_Σ is the *collective apparent power*, P_Σ is the *collective active power*, and Q_Σ is the *collective nonactive power*. The theory behind these expressions [17,18] is explained as follows: Each phase current i_k, $k = A, B, C, N$, is separated into two orthogonal components with reference to the voltage v_{kO}, see Fig. 5.7b, (the neutral path is considered a fourth phase if four-wire systems are involved),

$$i_k = i_{k\|} + i_{k\perp}$$

where $i_{k\|} = G_k v_{kO}$ is the *proportional component* and the *equivalent conductance*

$$G_k = P_k / V_{kO}^2$$

with P_k the total active power of phase k, carried by the voltage V_{kO}. The remaining current component is

$$i_{k\perp} = i_k - i_{k\|}$$

called the *orthogonal current*.

The total active power P_Σ absorbed by the load yields also a *mean value active equivalent conductance*

$$G = P_\Sigma / V_\Sigma^2$$

When the observed load is perfectly compensated to $PF = 1$, the load and compensator are equivalent to k resistances, $R_\Sigma = 1/G$, connected to the common node O, (see Fig. 5.12c).

Usually the proportional component of current $i_{k\|}$ differs from phase to phase. The next step is to separate $i_{k\|}$ in an active current called *symmetric proportional current*,

$$i_{kp} = G v_{kO}$$

and an *asymmetric proportional current*

$$i_{k\|u} = (G_k - G)v_{kO}$$

Since the total active power $P_\Sigma = \sum_k G V_{kO}^2$, results that $\sum_k G_k V_{kO}^2 = 0$, thus the asymmetric proportional currents $i_{k\|u}$ carry no active power. The nonactive components $i_{k\perp}$ and $i_{k\|u}$ lead to a *nonactive current*

$$i_{kq} = i_{k\perp} + i_{k\|u}$$

It can be proved that these currents are orthogonal, thus

$$I_{kq}^2 = I_{k\perp}^2 + I_{k\|u}^2 \tag{6.36}$$

and after multiplying (6.36) with V_{kO}^2 one obtains the nonactive powers of phase k.

$$Q_{tot\ k}^2 = Q_{tot\ k\perp}^2 + Q_{tot\ k\|u}^2 \tag{6.37}$$

For each one of the four phases we have

$$i_k = i_{kp} + i_{kq} = i_{kp} + i_{k\perp} + i_{k\|u}$$

which leads to the key expression, the basic relation between the rms current components of phase k

$$I_k^2 = I_{kp}^2 + I_{kq}^2 = I_{kp}^2 + I_{k\perp}^2 + I_{k\|u}^2 \tag{6.38}$$

that can be extended to the collective value

$$I_\Sigma^2 = I_{\Sigma p}^2 + I_{\Sigma q}^2 = I_{\Sigma p}^2 + I_{\Sigma\perp}^2 + I_{\Sigma\|u}^2 \tag{6.39}$$

Multiplying (6.39) with V_Σ^2 yields the DIN 40110 powers squared

$$S_\Sigma^2 = (V_\Sigma I_\Sigma)^2 = (V_\Sigma I_{\Sigma p})^2 + (V_\Sigma I_{\Sigma q})^2 = (V_\Sigma I_{\Sigma p})^2 + (V_\Sigma I_{\Sigma\perp})^2 + (V_\Sigma I_{\Sigma\|u})^2$$

leading to the final expressions

$$S_\Sigma^2 = P_\Sigma^2 + Q_{tot\ \Sigma p}^2 = P_\Sigma^2 + Q_{tot\ \Sigma\perp}^2 + Q_{tot\ \Sigma\|u}^2 \tag{6.40}$$

We find a lot in common between this approach and the FBD-method, presented in section 5.5, however, one will notice that the use of balanced and unbalanced susceptance is avoided in the DIN 40110. The nonactive currents $I_{k\perp b}$ (5.71) and $I_{k\perp u}$ (5.73) are included in I_{kq}. It is also important to realize that this standard does not allocate separate powers caused by voltage and current components with frequencies different than 60/50 Hz.

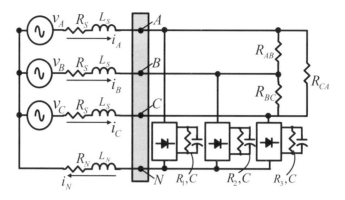

Figure 6.3   Example: The studied system.

EXAMPLE

In Fig. 6.3 is sketched the studied circuit; a three-phase symmetrical voltage 226.7 V rms, 60 Hz, supplies a four-wire soft line with the equivalent components $R_S = 0.1\ \Omega$, $R_N = 0.4\ \Omega$, $L_S = 1.2$ mH, $L_N = 2.0$ mH. The load is unbalanced and consists of three resistances $R_{AB} = 60\ \Omega$, $R_{BC} = 20\ \Omega$ and $R_{CA} = 30\ \Omega$, Δ-connected, as well as three full wave rectifiers, connected line-to-neutral. The dc load supplied by the three rectifiers are: $R_1 = 5\ \Omega$ on Phase A, $R_2 = 10\ \Omega$ on Phase B, and $R_3 = 7\ \Omega$ on Phase C. The filter capacitor $C = 2.0$ mF for all three units. This is a case where the voltage at the loads' terminals is excessively distorted. It was so chosen to help give a "magnified view" of the differences between the power definitions promoted by the two approaches.

The measured voltages and currents (obtained from numerical simulation) are listed in Tables 6.1, 6.2, and 6.3.

The measured active powers are separated in three categories:

1. The Δ-connected resistances dissipate a 60 Hz active power $P_{\Delta 1} = 1399.16$ W and a non-60 Hz active power $P_{\Delta H} = 89.36$ W. Resulting in a total $P_\Delta = 14081$ W.
2. The rectifiers absorb a 60 Hz active power $P_{Y1} = 27378$ W and generate harmonic active power $P_{YH} = -460$ W, out of which 89 W flow to the Δ-connected resistances and the remaining 371 W are dissipated in the line resistances R_S and neutral R_N.

Table 6.1   Measured line-to-neutral voltage phasors (in V rms)

h	AN	BN	CN
1	$202.23\angle -10.11°$	$229.74\angle -126.60°$	$210.72\angle 114.90°$
3	$65.08\angle -7.85°$	$57.95\angle -5.63°$	$55.26\angle 5.82°$
5	$25.48\angle -7.00°$	$15.31\angle 136.80°$	$18.09\angle -106.40°$
7	$9.86\angle 21.15°$	$9.92\angle -23.41°$	$6.24\angle -149.6°$
9	$9.04\angle 58.80°$	$10.42\angle 111.20°$	$8.07\angle -132.80°$
rms	215.17	237.96	219.95
$\%THD_V$	36.07	27.30	28.38

Table 6.2 Measured line-to-line voltage phasors (in V rms)

h	AB	BC	CA
1	$377.55\angle 23.92°$	$379.01\angle -97.23°$	$366.92\angle 141.0°$
3	$7.52\angle -25.29°$	$11.60\angle -76.56°$	$17.33\angle 123.20°$
5	$38.91\angle -20.40°$	$28.48\angle 102.3°$	$3.57\angle -154.90°$
7	$7.50\angle 89.24°$	$14.52\angle -3.09°$	$16.06\angle -155.30°$
9	$8.68\angle -13.19°$	$4.16\angle 65.71°$	$10.32\angle -169.90°$
rms	370.49	380.41	369.77
$\%THD_V$	12.73	10.22	12.59

3. The total input active power, measured at the terminals A, B, C, and N, is $P = 40999$ W with the 60 Hz active power $P_1 = 41369$ W and the harmonic active power $P_H = -371$ W.

6.3.1 The IEEE Std. 1459–2010 Approach

For comparison purpose the effective voltages were computed using different expressions. The results are tabulated (see Table 6.4) in the following order: First for expression (5.109) with $\xi = P_\Delta/P_Y = 0.523$; next for $\xi = 1.0$, (5.112), Case III section 5.5.4; followed by Case I, $\xi = 0$, (5.110) and last Case II, $\xi \to \infty$, (5.111).

Taking as a "yardstick" the effective voltage (5.112), with $\xi = 1.0$, recommended by the IEEE Std. 1459–2010, one finds that other expressions yield differences that affect the value of the effective apparent power.

The computed effective current with the ratio $\rho = R_N/R_S = 4.0$ is $I_{e_{\rho=4}} = 67.68$ A and with $\rho = 1.0$ is $I_{e_{\rho=1}} = 68.62$ A. Taking $\xi = 0.523$ the two corresponding effective apparent powers and power factors are

$$S_{e_{\rho=1\xi=0.523}} = 45610 \text{ VA} \quad \text{yielding a power factor} \quad PF_{e_{\rho=1}} = P/S_{e_{\rho=1}} = 0.899$$
$$S_{e_{\rho=4\xi=0.523}} = 48560 \text{ VA} \quad \text{yielding a power factor} \quad PF_{e_{\rho=4}} = P/S_{e_{\rho=4}} = 0.844$$

As explained in section 5.5.3, taking $\rho = 1$ leads to a better power factor. The 60 Hz or the fundamental effective current is $I_{e1} = 66.42$ A and the voltage $V_{e1} = 214.42$ V, leading to fundamental effective apparent power $S_{e1} = 3V_{e1}I_{e1} = 42725.84$ VA.

Table 6.3 Measured line current phasors (in A rms)

h	A	B	C	N
1	$73.04\angle 23.92°$	$53.80\angle -97.23°$	$69.54\angle 141.0°$	$12.97\angle -157.79°$
3	$11.86\angle -25.29°$	$6.40\angle -76.56°$	$7.78\angle 123.20°$	$21.51\angle 96.11°$
5	$10.08\angle -20.40°$	$8.42\angle 102.3°$	$7.27\angle -154.90°$	$0.99\angle 47.41°$
7	$2.29\angle 89.24°$	$1.97\angle -3.09°$	$2.91\angle -155.30°$	$0.70\angle 178.22°$
9	$1.48\angle -13.19°$	$1.00\angle 65.71°$	$1.04\angle -169.90°$	$0.97\angle -166.50°$
rms	74.61	54.61	70.35	25.18
$\%THD_I$	21.82	20.45	16.10	166.37

Table 6.4 Computed Effective Voltages (in V rms)

Expression	(5.109) or (6.2)	(5.112)	(5.110)	(5.111)
	221.56	220.18	224.57	215.69
% diference	0.63	0.0	1.99	-2.04

The non 60 Hz apparent power is conveniently derived

$$S_{eN} = \sqrt{S_{e_{\rho=1,\xi=1}}^2 - S_{e1}^2} = 15128.38 \text{ VA}$$

The values of harmonic effective current (6.19) and harmonic effective voltage (6.24), $I_{eH} = 17.24$ A and $V_{eH} = 49.00$ V, respectively, help compute the equivalent total harmonic distortions

$$THD_{eI} = \frac{I_{eH}}{I_{e1}} = 0.259 \quad \text{and} \quad THD_{eV} = \frac{V_{eH}}{V_{e1}} = 0.233$$

The remaining components of S_e can now be found

$$D_{eI} = (THD_{eI})S_{e1} = 3925.88 \text{ var}$$
$$D_{eV} = (THD_{eV})S_{e1} = 3527.51 \text{ var}$$
$$S_{eH} = (THD_{eI})S_{e1} = 915.40 \text{ VA}$$
$$D_{eH} = \sqrt{S_{eH}^2 - P_H^2} = 790.96 \text{ var}$$

The 60 Hz positive-sequence components are found using the positive-sequence voltage and current phasors, $V_1^+ = 212.40 - j26.88$ V and $I_1^+ = 63.04 - j16.25$ A, that give $P_1^+ = 38857.10$ W, $Q_1^+ = -15434.90$ var, $S_1^+ = 41810.40$ VA and $PF_1^+ = P_1^+/S_1^+ = 0.929$ The fundamental unbalanced power is

$$S_{U1} = \sqrt{S_{e1}^2 - (S_1^+)^2} = 8797.0 \text{ VA}$$

The ratio $S_{U1}/S_{e1} = 0.206$ is a preliminary evaluation of the degree of imbalance.

6.3.2   The DIN 40110 Approach

We start by determining, based on simulation, the rms voltages line-to-virtual neutral point O,

$$V_{A0} = 210.901V \quad V_{B0} = 221.638V \quad V_{C0} = 216.476V \quad V_{N0} = 46.403V$$

The collective voltage is

$$V_\Sigma = \sqrt{V_{AO}^2 + V_{BO}^2 + V_{CO}^2 + V_{NO}^2} = 374.68 \text{ V}$$

The measured phase active powers are

$$P_A = 14478 \text{ W} \quad P_B = 12314 \text{ W} \quad P_C = 14210 \text{ W} \quad P_N = 0 \text{ W}$$

with the collective active power $P_\Sigma = 41002$ W.

The knowledge of these values enable the computation of the equivalent conductances:

$$G_A = P_A/V_{AO}^2 = 0.326 \quad G_B = P_B/V_{BO}^2 = 0.251 \quad G_C = P_C/V_{CO}^2 = 0.303 \quad G_N = 0$$

and the mean value active equivalent conductance $G = P_\Sigma/V_\Sigma^2$.

The asymmetrical proportional currents are

$$I_{A\|u} = |G_A - G|V_{AO} = 3.36 \text{ A} \quad I_{B\|u} = |G_B - G|V_{BO} = 8.20 \text{ A}$$
$$I_{C\|u} = |G_C - G|V_{CO} = 8.01 \text{ A} \quad I_{N\|u} = |G_N - G|V_{NO} = 13.35 \text{ A}$$

yielding the collective asymmetrical proportional current:

$$I_{\Sigma\|u} = \sqrt{I_{A\|u}^2 + I_{B\|u}^2 + I_{C\|u}^2 + I_{N\|u}^2} = 17.92 \text{ A}$$

The nonactive component of powers are as follows: the proportional asymmetric nonactive power

$$Q_{tot\ \Sigma u} = V_\Sigma I_{\Sigma\|u} = 6764.19 \text{ var}$$

The total nonactive power

$$Q_{tot\ \Sigma} = \sqrt{S_\Sigma^2 - P_\Sigma^2} = 18221.44 \text{ var}$$

and the orthogonal nonactive power

$$Q_{tot\ \Sigma\perp} = \sqrt{Q_{tot\ \Sigma}^2 - Q_{tot\ \Sigma u}^2} = 16919.41 \text{ var}$$

The resulting power factor is $PF_\Sigma = P_\Sigma/S_\Sigma = 0.9138$.

In spite of the extreme case used in this example (large voltage distortion) and the quite different conceptual approaches, the results are quite close $PF_\Sigma \approx PF_{e_{\rho=1}}$. However, the IEEE method emphasizes the importance of the positive-sequence fundamental powers and provides quantitative information about the nonactive powers due to current distortion separated from voltage distortion. The DIN approach avoids the separation of fundamental components from the total apparent power. If power factor compensation is used to bring $PF \approx 1.0$, then for this particular example, where only one linear load exists, both methods will give identical and perfect results.

6.4 Observations and Suggestions

It is clear that the complete characterization and quantification of the power flow in polyphase systems involves more terms than a hands-on engineer would love to have. Moreover, the present power definitions are still evolving and related concepts did not receive universal approval. Following is presented a discussion meant to stimulate readers and students interested in energy generation, flow, and monitoring, to join the ongoing efforts and think of improvements to the present power definitions.

The present standards, DIN 40110 and IEEE Std. 1459–2010 recommend the same expression for the effective currents $I_\Sigma/\sqrt{3}$ and I_e. Nevertheless, one may address the following practical aspect: The rms currents I_a, I_b, I_c, and I_n cause the conductors' heating. In overhead lines the heat transfer conditions for one conductor are not affected by the currents flowing in the other lines, hence it does not seem correct to transfer the neutral path losses to the three phases. This observation leads to power loss equivalence that involves only the three phases:

$$\Delta P = 3R_S I_e^2 = R_S(I_a^2 + I_b^2 + I_c^2)$$

giving the four-wire system the same effective current as for the three-wire system:

$$I_e = \sqrt{\frac{I_a^2 + I_b^2 + I_c^2}{3}}$$

For a three-phase cable, or for bundles or trays with cables, the neutral current power losses affect the heat transfer conditions of the entire cable and the term ρI_n^2, that weighs the neutral current caused power loss, cannot be ignored when the effects of thermal stress are accounted for.

The true "soul searching" issue remains the definition of the three-phase voltages that yield Ve or V_Σ. The IEEE Std. 1459–2010 advocates a positive-sequence voltage V_e. DIN 40110 assumes no voltage changes at the terminals of the observed load. True, the consumers have very limited control over the quality of the supplied voltage, (unless a voltage regulator is installed and operated by the consumer). The original voltages are characterized, before the compensation, by certain spectra, unbalance, and harmonic phasors. After the compensation to unity power factor these characteristics change. The distortion and the asymmetry are reduced and the fundamental load voltages increase. It is not practical to assume that voltages remain unchanged after the total or partial power factor compensation.

The apparent power is measured or calculated in order to evaluate or predict insulation thermal stress, equipment aging, determine loading of equipment, establish penalties or bonuses, and compare performances among different converters or rotating machines. However, these are not activities that necessarily end with the actual implementation of the load compensation to unity power factor. The apparent power is a defined, convention type, quantity, that represents an ideal objective, a condition that is of advantage, economical and technical, for all end-users supplied by a power grid as well as for the owners and operators of the grid. In real life such an ideal situation cannot be obtained unless both the user and the providers of electric energy cooperate to reduce voltage and current waveforms distortion and help to clean the electromagnetic pollution, installing harmonic filters and power conditioners. The ultimate goal is to create an ideal power grid with positive-sequence voltages and currents only. This is the approach advocated by the IEEE Std. 1459–2010. The question that remains to be addressed is what is the right value of V_e?

We start by assuming a three-phase induction motor supplied with nonsinusoidal and unbalanced voltages [19]. Ignoring the skin effect the mechanical torque delivered by the motor during steady-state conditions is

$$T = T_1^+ + T_1^- + T_H^+ + T_H^- - \Delta T$$

where

$$T_1^+ = \frac{3(V_1^+)^2 R_2'/s}{\omega_S[(R_1 + R_2'/s)^2 + X^2]}$$

is the fundamental, positive-sequence torque, the main and only useful torque. The remaining torques are detrimental to the process of electromechanical energy conversion.

$$T_1^- = \frac{-3(V_1^+)^2 R_2'/(2-s)}{\omega_S\{[R_1 + R_2'/(2-s)]^2 + X^2\}}$$

is the fundamental, negative sequence torque, $T_1^- < 0$. It is a braking torque and the negative-sequence current causes significant additional losses.

$$T_H^+ = \sum_{h^+} \frac{R_2'/(s_{h^+})}{\omega_S\{[R_1 + R_2'/(s_{h^+})]^2 + X^2\}}$$

is the sum of the positive-sequence harmonic torques, and

$$T_H^- = \sum_{h^-} \frac{R_2'/(s_{h^-})}{\omega_S\{[R_1 + R_2'/(s_{h^-})]^2 + X^2\}}$$

is the sum of the negative-sequence harmonic torques, where the slip $s = 1 - \omega_m/\omega_S$ and ω_m is the rotor mechanical angular velocity, rad/s and ω_s is the synchronous angular velocity of the fundamental rotating field).

V_1^+ and V_1^- are the fundamental positive- and negative-sequence voltages, respectively, V_h^+ and V_h^- are the positive- and negative-sequence harmonic voltages, respectively, h^+ and h^- are the harmonic orders for the the positive- and negative-sequence harmonics.

The slips that correspond to the rotating fields of harmonics are $s_{h^+} = 1 - (1-s)/h^+ \approx 1.0$ for the positive-sequence harmonics and $s_{h^-} = 1 + (1-s)/h^- \approx 1.0$ for the negative-sequence harmonics.

One shall keep in mind that the interaction among the rotating fields of harmonic and the rotor harmonic currents, causes parasitic torques and vibrations,

R_1, R_2', and X are per phase values of stator resistance, rotor resistance (reflected to the stator) and the total leakage reactance.

ΔT represents the torque that covers the mechanical losses (windage and bearings).

If we supply this induction motor with a positive-sequence voltage V_e, that will cause the same output power such voltage as to satisfy the equation

$$T = \frac{3(V_e)^2 R_2'/s}{\omega_S[(R_1 + R_2'/s)^2 + X^2]} + \Delta T$$

thus

$$V_e = \sqrt{\frac{1}{3R'/s}\omega_S[(R_1 + R_2'/s)^2 + X^2](T - \Delta T)}$$

will ensure the same mechanical output, torque, and velocity, as in the original situation. It makes good sense to use in this case the voltage V_e as the correct voltage for computation of $S_e = 3V_e I_e$. The voltage $V_e < V_1^+$.

Next we shall discuss a three-phase rectifier. We consider a three-pulse midpoint connection rectifier (simpler to analyze than a six-pulse rectifier). We assume the following unbalanced voltages:

$$v_a = \widehat{V} \sin(\omega t)$$
$$v_b = b\widehat{V} \sin(\omega t - \beta)$$
$$v_c = c\widehat{V} \sin(\omega t - \gamma)$$

The conditions $v_a = v_b$, $v_b = v_c$ and $v_c = v_a$ provide the characteristic points of the envelope of direct voltage v_{dc}. These intersection points are found at the angles

$$\alpha_{ab} = \tan^{-1}\left[\frac{-\sin(\beta)}{1 - b\cos(\beta)}\right] ; \qquad \beta < \alpha_{ab} < \pi$$

$$\alpha_{bc} = \tan^{-1}\left[\frac{b\sin(\beta) - c\sin(\gamma)}{b\cos(\beta) - \cos(\gamma)}\right] ; \quad \gamma < \alpha_{bc} < \beta + \pi$$

$$\alpha_{ca} = \tan^{-1}\left[\frac{-c\sin(\gamma)}{1 - c\cos(\gamma)}\right] ; \qquad 2\pi < \alpha_{ca} < \gamma + \pi$$

thus the equivalence of direct voltage will be written as

$$V_{dc} = \frac{\widehat{V}}{2\pi}\left[\int_{\alpha_{ab}}^{\alpha_{bc}} b\sin(\omega t - \beta)d(\omega t) + \int_{\alpha_{bc}}^{\alpha_{ca}} c\sin(\omega t - \gamma)d(\omega t) + \int_{\alpha_{ca}}^{\alpha_{ab}+\pi} \sin(\omega t)d(\omega t)\right]$$

$$= \frac{3\sqrt{3}}{\pi\sqrt{2}}V_e = 1.17V_e$$

In this case the value of the equivalent V_e depends on the rectifier type, effect of commutation, filtering system, and the coefficients b and c, as well as the deviation from 120° (i.e. 120° − β and 120° + γ). Rectifiers sensitive to peak voltage inject noncharacteristic harmonics and their performance is strongly affected by the quality of voltage [20].

While it is possible for individual loads to determine (measure) an equivalent positive-sequence voltage, this task becomes impossible for loads that are mixed. We need a voltage that fits all. The answer is not obvious, and compromises are needed, but the agreement rules for such decisions are not yet available. One may consider using the fundamental positive–sequence voltage measured at the point of common coupling, or even to agree on a voltage that is the arithmetic or geometric mean of the weighed rated voltages of all the loads (according to the VA of each load).

Until new advances on the definitions and resolution of powers are recognized, the author will favor the IEEE approach where the voltage V_e determination is based on the evaluation of the active power supplied to the load.

6.5   Problems

6.1 A three-phase, four-wire customer is supplied with the following line-to-neutral harmonic voltage phasors (in V rms):

Table 6.5   Line-to-neutral voltage phasors

h	AN	BN	CN
1	$220.0\angle0.0°$	$210.0\angle-126.0°$	$225.0\angle114.0°$
3	$10.0\angle-5.0°$	$7.0\angle-8.0°$	$8.0\angle6.0°$
5	$5.0\angle-17.0°$	$4.0\angle130.0°$	$4.0\angle-106.0°$
7	$4.0\angle30.0°$	$2.0\angle-76.0°$	$3.0\angle150.0°$
9	$3.0\angle60.0°$	$2.0\angle84.0°$	$3.0\angle90.0°$

Table 6.6   Line current phasors (A rms)

h	A	B	C
1	$100.0\angle-30.0°$	$105.0\angle-126.0°$	$95.0\angle114.0°$
3	$16.0\angle-88°$	$14.0\angle-93.0°$	$12.0\angle-87.0°$
5	$9.0\angle-105.0°$	$13.0\angle39.0°$	$15.0\angle-193.0°$
7	$1.0\angle-58.0°$	$1.0\angle-76.0°$	$3.0\angle150.0°$
9	$1.0\angle-28.0°$	$2.0\angle-5.0°$	$3.0\angle1.0°$

Your task is to compute the powers defined by the three methods explained in this chapter and to compare the results. Assume $\rho = 1.0$ and $\xi = 1.0$. The line-to-line voltages can be determined from the line-to-neutral voltages and the neutral current from $i_N = -(i_A + i_B + i_C)$.

6.2 Work the example given in this chapter (Fig. 6.3). Include more information such as vector apparent power.

6.3 A three-phase, three-wire line supplies a customer with perfectly symmetrical line-to-line 660 V rms, 60 Hz. The load consists of two resistances, $R_{AB} = 6\ \Omega$ and $R_{BC} = 36\ \Omega$, connected line-to-line, A–B and B–C, respectively. There is also a large single-phase rectifier connected A–to–C. The equivalent load of this rectifier is a current source $I_{dc} = 100$ A. Your task is to test different power definitions applied to this circuit. An interesting aspect to be discussed is the fact that this is a three-wire system and many people, incorrectly, expect no third harmonic to be present in the supplying line. When you compute the current harmonics you will find that triplen harmonics do flow. Explain the reason.

6.4 Repeat problem 6.3 if the line-to-line source voltage is increased to 710 V and the feeder, assumed well balanced, has an equivalent per phase resistance $R_S = 0.05\ \Omega$ and inductance $L_S = 1.5$ mH.

6.5 Two industrial customers are supplied with a three-phase, four wire system, 14 kV, 60 Hz, by a feeder with $R_S = 0.2\ \Omega$ and $L_S = 4.0$ mH (about 5kA symmetrical short-circuit current). The neutral path, from substation to customers' transformers, has the equivalent components $R_N = 0.3\ \Omega$ and $L_N = 5.0$ mH. The first customer is an uncontrolled six-pulse rectifier supplying a dc load that can be simulated by a current source $I_{dc} = 150$ A. The rectifier is connected to the A, B, and C lines via three equivalent inductances $L_1 = 0.5$ mH. The second customer has two loads: one is an identical rectifier as the one described and the second load is a Y-connected unbalanced load. The unbalanced load can be described as follows: $R_{AN} = 300\ \Omega$, $R_{BN} = 600\ \Omega$, $R_{CN} = 100\ \Omega$, $L_{AN} = 400$ mH, $L_{BN} = 990$ mH and $L_{CN} = 300$ mH. The second customer is connected to the A, B, and C lines via three equivalent inductances $L_2 = 0.5$ mH. The Y-connected load has a neutral path with $R_{NN} = 5$ mΩ and $L_{NN} = 0.03$ mH.

Your task is to compare the IEEE and DIN approaches when the second customer corrects the *PF* such that it becomes ≈ 1.0. You should focus on current and voltage distortion and symmetrical components (see section 5.5.4 for a numerical example).

6.6   References

[1] Shepherd W., Zand, P.: "Energy Flow and Power Factor in Non-sinusoidal Circuits," Cambridge University Press, New York, 1979.

[2] Arrillaga J., Bradley, D. A., Bodger, P. S.: "Power Systems Harmonics," John Wiley and Sons, 1985, pp. 122–23.

[3] Kassakian J. K., Schlecht, M. F., Verghese, G. C.: "Principles of Power Electronics," Addison-Wesley Publishing Co., 1991, pp. 45–52.

[4] Emanuel A. E.: "Powers in Nonsinusoidal Situations–A Review of Definitions and Physical Meaning," IEEE Trans. on Power Delivery, Vol. 3, No. 5, July 1990, pp. 1377–89.

[5] Czarnecki L. S.: "Orthogonal Decomposition of the Currents in a 3-Phase Non-linear Asymmetrical Circuit with Non-sinusoidal Voltage Source," IEEE Trans. on Instr. and Meas., Vol IM–37, No. 1, 1987, pp. 30–34.

[6] Czarnecki L. S.: "Scattered and Reactive Current, Voltage and Power in Circuits with Non-sinusoidal Waveforms and their Compensation," IEEE Trans. on Instr. and Meas., Vol IM–40, No. 3, June 1991, pp. 563–67.

[7] Czarnecki L. S.: "Minimization of Unbalanced and Reactive Currents in Three-Phase Asymmetrical Circuits with Non-sinusoidal Voltage," IEE Proceedings–B, Vol. 139, No. 4, July 1992, pp. 347–359.

[8] Depenbrock M.: "Some Remarks to Active and Fictitious Power in Polyphase and Single Phase Systems," ETEP, Vol. 3, Jan. 1962, pp. 15–19.

[9] Depenbrock M.: "The FBD, Method – a Generally Applicable Tool for Analyzing Power Relations," IEEE Trans. on Power Systems, Vol. 8, No. 2, May 1993,

[10] Ferrero A., Morando, A. P., Ottoboni, R., Superti-Furga, G.: "On the Meaning of the Park Power Components in Three-Phase Systems under Non-sinusoidal Conditions," ETEP, Vol. 3, No. 1, Jan./Feb. 1933, pp. 33–43.

[11] Ferrero A., Superti-Furga, G.: "A New Approach to the Definition of Power Components in Three-Phase Systems Under Non-sinusoidal Conditions," IEEE Tran. Instr. and Meas. Vol. 40, No. 3, June 1991, pp. 568–77.

[12] Malengret M., Trevor Gaunt, C.: "Decomposition of Currents in Three- and Four-Wire Systems," IEEE Trans. on Instr. and Meas. Vol. 57, No. 5, May 2008, pp. 963–72.

[13] Bullock D. F., Elmore, D. D.: "MinD Ur Ps & Qs. Four Powers, Computed from Voltages and Currents under IEEE Std–100. Meet Rational Metering Requirements," GE Meter Business. Somerworth, NH 03878.

[14] Curtis H. L., Silsbee, F. B.: "Definitions of Power and Related Quantities," Electrical Engineering, April 1935, pp. 394–404.

[15] American Standard Definitions of Electrical Terms, August 12, 1941, pp. 35–46. (Sponsored by AIEE, Approved by American Standards Association and Canadian Engineering Standards Association.)

[16] IEEE 100. The Authoritative Dictionary of IEEE Standards Terms. Seventh Edition. IEEE Standards Information Network. 2000. IEEE Standards Dictionary: Glossary of Terms and Definitions, 2009.

[17] DIN 40110–2:2002–11, Quantities Used in Alternating Current Theory–Part 2: Multi-Line Circuits.

[18] Spath H.: "A General Purpose Definition of Active Current and Non-Active Power Based on German Standard DIN 40110.

[19] Smolensky A. I., "Electrical Machines," Mir Publishers, Energy, Moscow, 1982.

[20] Emanuel A. E., Orr J. A.: "Six-Pulse Converters Atypical Harmonics Caused by Second Harmonic Voltage," Tenth International Conference on Harmonics and Quality of Power, Rio de Janeiro, Brazil, October 2002, IEEE Catalog No. 02EX630C.

7

Power Definitions for Time-Varying Loads

An idea starts by being a paradox. Continues by becoming a banality, and ends by being a prejudice.

—Grigore C. Moisil, *Mathematician*

The loads supplied by an electric power system are varying continuously in time. Some time variations are gradual, presenting a slow, continuous trend characterized by a nearly constant rate of change of powers. Such a situation is typical for large clusters of loads. Other time-variations of power are more abrupt: lights are turned on and off, motors driving machine tools start and stop, welders and compressors sometimes operate repetitively, sometimes randomly. Elevators and electric cranes have a definite probabilistic behavior. Arc furnaces in the initial stage, when the charge is a conglomerate of metallic masses (scrap), present a time-varying load supplied with a distorted current whose waveform is not repetitive even from one cycle to the next; quite incorrectly, some people consider that such nonrepetitive currents shall be represented by equivalent time-varying harmonics.

When power measurements are taken over a time duration τ (seconds, minutes, hours, or days), the observation time τ is divided in ν equal subintervals $\Delta\tau$, where $\tau = \nu\Delta\tau$. Instruments that implement the real time data acquisition for each time interval $1 \leq i \leq \nu$, are usually measuring the following quantities:

- harmonic voltage phasors: $V_{hi}\angle\alpha_{hi}$, h=1, 2, 3...;
- harmonic current phasors: $I_{hi}\angle\beta_{hi}$;
- interharmonic voltage and current phasors, (h is a noninteger number);
- total harmonic distortion of voltage and current;
- rms voltage and currents: V_i, I_i;
- Voltage and current symmetrical components: Positive-sequence: $V_{1i}^{+}\angle\alpha_{1i}^{+}$ and $I_{1i}^{+}\angle\beta_{1i}^{+}$. Negative-sequence: $V_{1i}^{-}\angle\alpha_{1i}^{-}$ and $I_{1i}^{-}\angle\alpha_{1i}^{-}$. Zero-sequence: $V_{1i}^{0}\angle\alpha_{1i}^{0}$ and $I_{1i}^{0}\angle\beta_{1i}^{0}$.

Based on this data, the apparent power S_i, active power P_i, and the nonactive powers N_i, Q_i, D_{Ii}, D_{Vi}, and the harmonic apparent power S_{Hi}, may be computed and recorded for

Power Definitions and the Physical Mechanism of Power Flow   Alexander Eigeles Emanuel
© 2010 John Wiley & Sons, Ltd

each time interval $\Delta \tau$[1]. It is recommended to design power and energy meters with $\Delta \tau = 200$ ms (ten cycles for 50 Hz power systems and twelve cycles for 60 Hz)[1]. The 200 ms values[2] are aggregated over additional intervals such as 3s, 10 min, or 2.0 h. The "aggregations are performed using the square root of the arithmetic mean of the squared input values" [1]. In some applications the mean values for each $\Delta \tau$ are recorded. For example, the Demand Meters measure the maximum mean power (kW, kVA, or kvar) over a time interval $\tau = 15$, 30, or 60 min and typically record only the maximum mean value for an observed month [2]. The goal of this chapter is to provide additional information on the best interpretation of the collected data.

The real life probabilistic nature of voltages, currents, and powers, is proved by actual measurements taken at a 15 kV substation that supplies a mixture of industrial and residential customers [3, 4], (Figs. 7.1 to 7.3). The mean values were recorded over an observation time $\tau = 168$ h $= 7$ day with $\Delta \tau = 10$ min. The rms voltage (Fig. 7.1a), has the probability distribution shown in (Fig. 7.1b), and resembles a Gaussian distribution. The weekly variations do not follow a specific visible pattern; this is due to voltage regulation by means of the tap changer at the substation transformer. The supplied current (Fig. 7.2 a), follows a daily pattern with a trough during the low demand hours and lower weekend values. The probability distribution of the current (Fig. 7.2a), has a visible multi modal distribution. The measured active power (Fig. 7.3a), follows a time-variation pattern similar to the current. The reactive power (Fig. 7.3b), is strongly affected by switched capacitors that are energized during the high demand times and cause power factor overcompensation.

7.1 Background: Basic Example

Let us assume an industrial oven with adjustable heating elements, which is supplied by a soft line. The measured rms voltage and current and the computed active power are summarized in Table 7.1. Five equal time intervals are observed. The rms voltage for the entire time $\tau = 5\Delta \tau$ is

$$V = \sqrt{\frac{1}{5} \sum_{i=1}^{5} V_i^2} = \sqrt{\frac{1}{5}(220^2 + 225^2 + 215^2 + 210^2 + 230^2)} = 220.11 \text{ V}$$

Similarly we obtain the rms current

$$I = \sqrt{\frac{1}{5} \sum_{i=1}^{5} I_i^2} = \sqrt{\frac{1}{5}(100^2 + 60^2 + 120^2 + 150^2 + 30^2)} = 101.39 \text{ A}$$

---

[1] The measurements during the subinterval $\Delta \tau$ are based on the assumption that the waveforms in the interval i are repeated during the next intervals $i + 1$, $i + 2$, ...

[2] When this book was prepared, instrumentation experts were debating if $\Delta \tau = 320$ ms for 50 Hz and $\Delta \tau = 267$ ms for 60 Hz should not replace the $\Delta \tau = 200$ ms.

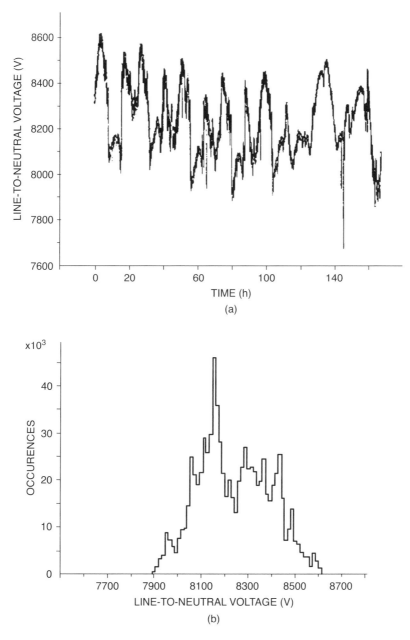

Figure 7.1 Line-to-neutral voltage: (a) Seven days time variations. (b) Relative frequency distribution. Source: A. E. Emanuel, J. A. Orr, D. Cyganski, E. M. Gulachenski, "A Survey of Harmonic Voltages and Currents at Distribution Substations," IEEE Trans. On Power delivery, Vol.6, No.4, Oct. 1991, pp.1883–1890.

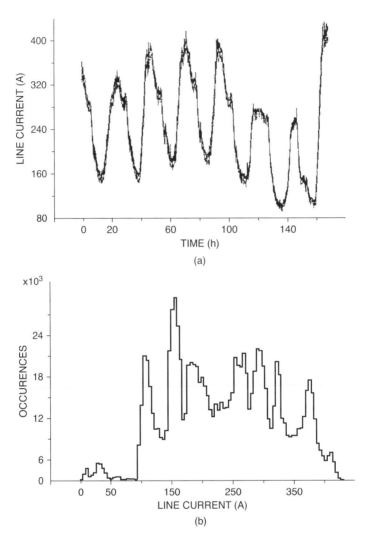

Figure 7.2 Line current: (a) Seven days time variations. (b) Relative frequency distribution. Source: A. E. Emanuel, J. A. Orr, D. Cyganski, E. M. Gulachenski, "A Survey of Harmonic Voltages and Currents at Distribution Substations," IEEE Trans. On Power delivery, Vol.6, No.4, Oct. 1991, pp.1883–1890.

Lastly is computed the power that will carry the same amount of energy, the active power:

$$P = \frac{1}{\tau} \int_0^\tau VI \, dt = \frac{1}{5} \sum_{i=1}^5 V_i I_i = \frac{1}{5}(22.00 + 13.50 + 25.80 + 31.50 + 6.60) = 19.88 \text{ kW}$$

and the apparent power

$$S = VI = 22.32 \text{ kVA}$$

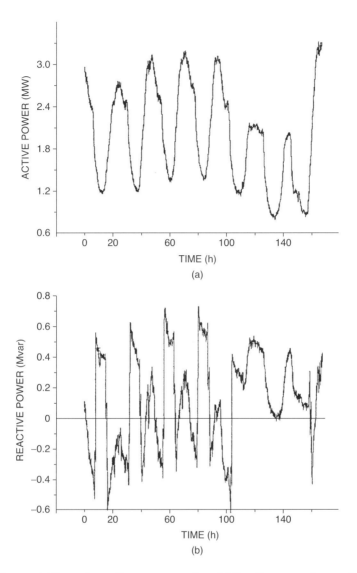

Figure 7.3 Powers: (a) Seven days active power recording. (b) Reactive power. Source: A. E. Emanuel, J. A. Orr, D. Cyganski, E. M. Gulachenski, "A Survey of Harmonic Voltages and Currents at Distribution Substations," IEEE Trans. On Power delivery, Vol.6, No.4, Oct. 1991, pp.1883–1890.

One realizes that $P < S$ in spite of the purely resistive nature of the load and the total lack of inductive or capacitive components. The straight explanation is that such a load behaves over the observation time τ like a nonlinear resistance, more correctly a time-varying resistance. Based on this observation it is possible to promote a new concept, a paradox, the existence of a nonactive power, *the Randomness Power*:

$$D_R = \sqrt{S^2 - P^2} = 10.14 \text{ kvar}$$

Table 7.1 Single-phase load with unity PF and randomly varying voltage and current (Basic Example)

i	1	2	3	4	5
V_i (V)	220	225	215	210	230
I_i (A)	100	60	120	150	30
P_i (kW)	22.00	13.50	25.80	31.50	6.60

responsible for a sub unity power factor

$$PF = P/S = 0.891$$

7.2   Single-Phase Sinusoidal Case

In this section the results observed in the previous basic example are generalized. We start from the active and reactive powers expressions measured during the i-interval:

$$P_i = V_i I_i \cos(\theta_i) ; \quad \theta_i = \alpha_i - \beta_i \tag{7.1}$$

$$Q_i = V_i I_i \sin(\theta_i) \tag{7.2}$$

The rms values of voltage and current for the entire observation time τ are

$$V = \sqrt{\frac{1}{\tau} \sum_{i=1}^{\nu} V_i^2} \qquad I = \sqrt{\frac{1}{\tau} \sum_{i=1}^{\nu} I_i^2} \tag{7.3}$$

leading to the apparent power expression:

$$S = VI = \sqrt{\frac{1}{\nu^2} \sum_{i=1}^{\nu} V_i^2 \sum_{i=1}^{\nu} I_i^2} \tag{7.4}$$

The active power for the duration τ is

$$P = \frac{1}{\tau} \sum_{i=1}^{\nu} P_i = \frac{1}{\nu} \sum_{i=1}^{\nu} V_i I_i \cos(\theta_i) = \bar{P} \tag{7.5}$$

and it is equal to the mean power $\bar{P}$. Comparing (7.4) with (7.5) one concludes that $S \geq \bar{P}$.

Conventional var meters are designed to operate with a 90° delaying of the voltage waveform with respect to the current waveform [5]. This concept is upheld in the IEEE Std. 1459–2010

[6] where for sinusoidal conditions the reactive power can be accurately measured using instruments based on the equation

$$Q = \frac{\omega}{\tau} \int_0^\tau i \left[\int v \, dt \right] dt \tag{7.6}$$

where τ is an integer number of cycles and $\omega = 2\pi f$ is the angular frequency.
    Substitution of

$$i_i = \sqrt{2} I_i \sin(\omega t + \beta) \quad \text{and} \quad v_i = \sqrt{2} V_i \sin(\omega t + \alpha)$$

in (7.6) leads to the value of reactive power for the time τ:

$$Q = \frac{1}{\nu} \sum_{i=1}^{\nu} V_i I_i \sin(\theta_i) = \bar{Q} \; ; \quad \theta_i = \alpha_i - \beta_i \tag{7.7}$$

The above result shows that the reactive power for the observation time τ is the mean value of the reactive powers measured for each subinterval $\Delta\tau$. The same conclusion is reached for a few other types of varmeters, except for the ones based on the expression

$$Q = \sqrt{S^2 - P^2}$$

In the previous chapters it was emphasized that the reactive power Q causes power losses in the conductors that supply the observed load, (2.32); This observation leads to a more rigorous definition of Q, a definition that ties the reactive power to ΔW, the energy lost in the supplying line. A practical approach proposed in [5] is based on the minimization of the energy lost on the account of Q_i:

$$\Delta W = r \sum_{i=1}^{\nu} \frac{P_i^2 + Q_i^2}{V_i^2} \Delta\tau \tag{7.8}$$

where r is the equivalent resistance of the supplying line.
    A linear capacitance connected in parallel to the load will help to minimize the energy lost. Such a capacitance will deliver the same optimum reactive power Q_c during each subinterval. Evidently, Q_c is a fictitious quantity that helps to understand the reactive power concept under time-varying conditions. The objective function to be minimized is

$$F(Q_c) = \sum_{i=1}^{\nu} \frac{(Q_i - Q_c)^2}{V_i^2} \tag{7.9}$$

Since $(Q_i - Q_c)^2 \geq 0$ there is not a value of Q_c that yields $F(Q_c) = 0$. Only when $Q_i = Q$ and $V_i = V$ are constant for the entire observation time, the obvious best value is $Q_c = Q_i = Q$. For the general case Q_c is found by solving the equation

$$\frac{dF(Q_c)}{dQ_c} = -\sum_{i=1}^{v} \frac{2(Q_i - Q_c)}{V_i^2} = -2\sum_{i=1}^{v} \frac{Q_i}{V_i^2} + 2Q_c \sum_{i=1}^{v} \frac{1}{V_i^2} = 0$$

that yields the quantity that fulfills the true physical meaning of reactive power:

$$Q_c = \frac{\displaystyle\sum_{i=1}^{v} \frac{Q_i}{V_i^2}}{\displaystyle\sum_{i=1}^{v} \frac{1}{V_i^2}} \tag{7.10}$$

This Q_c can be considered as the equivalent reactive power for the time τ. One can observe that

$$Q_c \neq \bar{Q} = \frac{1}{v}\sum_{i=1}^{v} Q_i \tag{7.11}$$

From (7.11) results that

$$Q_c = \bar{Q} + \Delta Q$$

and one learns that the randomness affects also the overall reactive power value: the difference

$$\Delta Q = Q_c - \bar{Q} = \frac{\displaystyle\sum_{i=1}^{v} \frac{Q_i}{V_i^2}}{\displaystyle\sum_{i=1}^{v} \frac{1}{V_i^2}} - \frac{1}{v}\sum_{i=1}^{v} Q_i$$

is a randomness component of the reactive power, and only when $V_i \approx V$ then $\Delta Q \approx 0$.

The following second example helps to clarify the above proposal. This time we consider an inductive load. In Table 7.2 are given the voltages V_i, currents I_i, their active and reactive current components, I_{pi}, I_{qi}, while the corresponding powers and power factor are provided for five equal time intervals.

The rms voltage and current are $V = 220.11$ V and $I = 101.39$ A, therefore the correct apparent power value is $S = VI = 22.32$ kVA. The mean value of the apparent power is $\bar{S} = 19.94$ kVA, about 10% less than the correct value (22.32 kVA).

The reactive power computations give $\bar{Q} = 12.77$ kvar and $Q_c = 13.02$ kvar. We find that in spite of relatively large excursions of the voltage V_i, the randomness component $\Delta Q = 0.25$ kvar is a relatively small quantity (1.9%).

It is also learned that the correct power factor $PF = P/S = 0.667$ and the mean power factor

$$\bar{PF} = \frac{1}{v}\sum_{i=1}^{v} \frac{P_i}{S_i} = 0.669$$

Table 7.2 Single-phase inductive load with randomly varying voltage
and current

i	1	2	3	4	5	MEAN VALUE
V_i (V)	220	225	215	210	230	220
I_i (A)	100	60	120	150	30	92
I_{pi} (A)	80	35	100	120	10	69
I_{qi} (A)	60	48.73	66.33	90.00	28.28	58.67
P_i (kW)	17.60	7.87	21.50	25.20	17.60	14.90
Q_i (kvar)	13.20	10.97	14.26	18.90	6.50	12.77
S_i (kVA)	22.00	13.50	22.00	31.50	6.90	19.94
PF_i	0.800	0.583	0.833	0.800	0.333	0.670

are almost equal. When the mean apparent power is used, the power factor $PF' = P/\bar{S} = 0.747$ gives the wrong information about the line utilization.

Finally we find the randomness power

$$D_R = \sqrt{S^2 - (P^2 + Q_c^2)} = 10.32 \text{ kvar}$$

If D_R is calculated using the mean value $\bar{Q}$ results

$$D_R = \sqrt{S^2 - [P^2 + (\bar{Q})^2]} = 10.63 \text{ kvar}$$

with a 3% difference caused by the randomness nature of ΔQ.

7.2.1 Analytical Expressions of Powers: Single-Phase Sinusoidal

For the observation time τ, the apparent power squared has the expression

$$S^2 = V^2 I^2 = \frac{1}{v^2} \sum_{i=1}^{v} V_i^2 \sum_{i=1}^{v} I_i^2$$

$$= \frac{1}{v^2} \sum_{i=1}^{v} V_i^2 \sum_{i=1}^{v} [I_i \cos(\theta_i)]^2 + \frac{1}{v^2} \sum_{i=1}^{v} V_i^2 \sum_{i=1}^{v} [I_i \sin(\theta_i)]^2$$

The apparent power resolution is obtained by using Lagrange's identity:

$$\sum_{i=1}^{v} a_i^2 \sum_{i=1}^{v} b_i^2 = \left(\sum_{i=1}^{v} a_i b_i\right)^2 + \sum_{1 \le n < m \le v} (a_m b_n - a_n b_m)^2$$

hence

$$S^2 = \frac{1}{v^2} \left(\sum_{i=1}^{v} V_i^2 I_i \cos(\theta_i) \right)^2 + \frac{1}{v^2} \sum_{1 \leq n < m \leq v} [V_m I_n \cos(\theta_n) - V_n I_m \cos(\theta_m)]^2$$

$$+ \frac{1}{v^2} \left(\sum_{i=1}^{v} V_i^2 I_i \sin(\theta_i) \right)^2 + \frac{1}{v^2} \sum_{1 \leq n < m \leq v} [V_m I_n \sin(\theta_n) - V_n I_m \sin(\theta_m)]^2$$

$$= (\bar{P})^2 + (\bar{Q})^2 + D_R^2$$

where the distortion power squared is

$$D_R^2 = \frac{1}{v^2} \sum_{1 \leq n < m \leq v} \left\{ [V_m I_n \cos(\theta_n) - V_n I_m \cos(\theta_m)]^2 + [V_m I_n \sin(\theta_n) - V_n I_m \sin(\theta_m)]^2 \right\}$$

$$= \frac{1}{v^2} \sum_{1 \leq n < m \leq v} [(V_m I_n)^2 + (V_n I_m)^2 - 2 V_m V_n I_m I_n \cos(\theta_m - \theta_n)] \qquad (7.12)$$

The practical way to compute the randomness power is

$$D_R = \sqrt{S^2 - (\bar{P})^2 - (\bar{Q})^2} \qquad (7.13)$$

Using $\bar{Q}$ instead of Q_c means that the component ΔQ is accounted for. This fact becomes clear if one substitutes $\bar{Q} = Q_c - \Delta Q$ in (7.13), resulting in

$$D_R = \sqrt{S^2 - P^2 - (\bar{Q})^2} = \sqrt{S^2 - (\bar{P})^2 - Q_c^2 \left(1 - \frac{\Delta Q}{Q_c} \right)^2} \qquad (7.14)$$

7.3   Single-Phase Nonsinusoidal Case

Every subinterval i is characterized by its rms voltage and current V_i, I_i, respectively. The two major components of V_i and I_i are the fundamental rms voltage and current, V_{1i}, I_{1i}, and the total rms harmonic voltage and current, V_{Hi}, I_{Hi}, i.e.,:

$$V_i^2 = V_{1i}^2 + V_{Hi}^2 \quad \text{and} \quad I_i^2 = I_{1i}^2 + I_{Hi}^2$$

The rms values of these components for the entire interval τ are

$$V_1 = \sqrt{\frac{1}{v} \sum_{i=1}^{v} V_{1i}^2} ; \quad I_1 = \sqrt{\frac{1}{v} \sum_{i=1}^{v} I_{1i}^2}$$

$$V_H = \sqrt{\frac{1}{v} \sum_{i=1}^{v} V_{Hi}^2} ; \quad I_H = \sqrt{\frac{1}{v} \sum_{i=1}^{v} I_{Hi}^2}$$

hence the apparent power squared has the following expressions

$$S^2 = (V_1^2 + V_H^2)(I_1^2 + I_H^2) = \frac{1}{v^2}\left(\sum_{i=1}^{v} V_{1i}^2 + \sum_{i=1}^{v} V_{Hi}^2\right)\left(\sum_{i=1}^{v} I_{1i}^2 + \sum_{i=1}^{v} I_{Hi}^2\right)$$

$$= \frac{1}{v^2}\sum_{i=1}^{v} V_{1i}^2 \sum_{i=1}^{v} I_{1i}^2 + \frac{1}{v^2}\sum_{i=1}^{v} V_{1i}^2 \sum_{i=1}^{v} I_{Hi}^2 + \frac{1}{v^2}\sum_{i=1}^{v} V_{Hi}^2 \sum_{i=1}^{v} I_{1i}^2 + \frac{1}{v^2}\sum_{i=1}^{v} V_{Hi}^2 \sum_{i=1}^{v} I_{Hi}^2$$

$$= S_1^2 + D_I^2 + D_V^2 + S_H^2 \tag{7.15}$$

with four distinct terms:

1. The first term is the fundamental apparent power squared

$$S_1^2 = \frac{1}{v^2}\left(\sum_{i=1}^{v} V_{1i}^2 I_{1i}\cos(\theta_{1i})\right)^2 + \frac{1}{v^2}\sum_{1 \leq n < m \leq v}[V_{1m}I_{1n}\cos(\theta_{1n}) - V_{1n}I_{1m}\cos(\theta_{1m})]^2$$

$$+ \frac{1}{v^2}\left(\sum_{i=1}^{v} V_{1i}^2 I_{1i}\sin(\theta_{1i})\right)^2 + \frac{1}{v^2}\sum_{1 \leq n < m \leq v}[V_{1m}I_{1n}\sin(\theta_{1n}) - V_{1n}I_{1m}\sin(\theta_{1m})]^2$$

$$= (\bar{P}_1)^2 + (\bar{Q}_1)^2 + (D_{1R})^2 \tag{7.16}$$

where

$$(D_{1R})^2 = \frac{1}{v^2}\sum_{1 \leq n < m \leq v}[(V_{1m}I_{1n})^2 + (V_{1n}I_{1m})^2 - 2V_{1m}V_{1n}I_{1m}I_{1n}\cos(\theta_{1m} - \theta_{1n})]$$

is the *Fundamental Randomness Power*. This result is identical to (7.12)

2. The second term is the current distortion power squared

$$D_I^2 = \frac{1}{v^2}\sum_{i=1}^{v} V_{1i}^2 \sum_{i=1}^{v} I_{Hi}^2 = \frac{1}{v^2}\left(\sum_{i=1}^{v} V_{1i}I_{Hi}\right)^2 + \frac{1}{v^2}\sum_{1 \leq n < m \leq v}(V_{1m}I_{1n} - V_{1n}I_{1m})^2$$

$$= (\bar{D}_I)^2 + (D_{IR})^2 \tag{7.17}$$

where

$$D_{IR} = \sqrt{D_I^2 - (\bar{D}_I)^2} = \sqrt{\frac{1}{v^2}\sum_{1 \leq n < m \leq v}(V_{1m}I_{1n} - V_{1n}I_{1m})^2} \tag{7.18}$$

is the *Current Distortion Randomness Power*.

3. The third term is the voltage distortion power squared

$$D_V^2 = \frac{1}{v^2} \sum_{i=1}^{v} V_{Hi}^2 \sum_{i=1}^{v} I_{1i}^2 = \frac{1}{v^2} \left(\sum_{i=1}^{v} V_{Hi} I_{1i} \right)^2 + \frac{1}{v^2} \sum_{1 \le n < m \le v} (V_{Hm} I_{1n} - V_{Hn} I_{1m})^2$$
$$= (\bar{D}_V)^2 + (D_{VR})^2 \tag{7.19}$$

where

$$D_{VR} = \sqrt{D_V^2 - (\bar{D}_V)^2} = \sqrt{\frac{1}{v^2} \sum_{1 \le n < m \le v} (V_{Hm} I_{1n} - V_{Hn} I_{1m})^2} \tag{7.20}$$

is the *Voltage Distortion Randomness Power*.
4. The last term is the harmonic apparent power squared

$$S_H^2 = \frac{1}{v^2} \sum_{i=1}^{v} V_{Hi}^2 \sum_{i=1}^{v} I_{Hi}^2 = \frac{1}{v^2} \left(\sum_{i=1}^{v} V_{Hi} I_{Hi} \right)^2 + \frac{1}{v^2} \sum_{1 \le n < m \le v} (V_{Hm} I_{Hn} - V_{Hn} I_{Hm})^2$$
$$= (\bar{S}_H)^2 + (S_{HR})^2 \tag{7.21}$$

where

$$S_{HR} = \sqrt{S_H^2 - (\bar{S}_H)^2} = \sqrt{\frac{1}{v^2} \sum_{1 \le n < m \le v} (V_{Hm} I_{Hn} - V_{Hn} I_{Hm})^2} \tag{7.22}$$

is the *Harmonic Apparent Randomness Power*.

The resolution of the apparent power is found in the following expression:

$$S^2 = (\bar{P}_1)^2 + (\bar{Q}_1)^2 + (\bar{D}_I)^2 + (\bar{D}_V)^2 + (\bar{S}_H)^2 + D_R^2 \tag{7.23}$$

where

$$D_R = \sqrt{(D_{1R})^2 + (D_{IR})^2 + (D_{VR})^2 + (S_{HR})^2}$$

is the *total Randomness Power*.

7.4   Three-Phase Sinusoidal and Unbalanced Condition

Three-phase systems require the measurement of the effective voltage V_e and current I_e. From (5.103) and (5.104) it is found

$$V_e^2 = (V^+)^2 + V_u^2 \quad \text{and} \quad I_e^2 = (I^+)^2 + I_u^2$$

where

$$V_u^2 = (V^-)^2 + \frac{(V^0)^2}{4} \quad \text{and} \quad I_u^2 = (I^-)^2 + 4(I^0)^2$$

are components caused by the system unbalance; the voltages V^+, V^-, and V^0 are the positive-, negative-, and zero-sequence line-to-line voltages, and the currents I^+, I^-, and I^0 are the positive-, negative-, and zero-sequence line currents. The effective apparent power squared measured for a subinterval $\Delta\tau$ is

$$S_{ei}^2 = (3V_{ei}I_{ei})^2 = (S_i^+)^2 + (S_{ui})^2 \tag{7.24}$$

where

$$S_i^+ = 3V_i^+ I_i^+ = \sqrt{(P_i^+)^2 + (Q_i^+)^2}$$

is the positive-sequence apparent power and

$$P_i^+ = 3V_i^+ I_i^+ \cos(\theta_i^+) \ ; \qquad Q_i^+ = 3V_i^+ I_i^+ \sin(\theta_i^+)$$

are the positive-sequence active and reactive powers. These three powers, S_i^+, P_i^+, and Q_i^+, are major components that dominate the flow of electric energy.

The unbalanced apparent power, S_{ui} in (7.24), has three terms

$$(S_{ui})^2 = 9[(V_{ui}I_{ui})^2 + (V^+ I_{ui})^2 + (V_{ui}I_i^+)^2] \tag{7.25}$$

The first term includes active and nonactive powers associated with the negative- and the zero-sequence currents and voltages:

$$(V_{ui}I_{ui})^2 = 9\left[(V_i^- I_i^-)^2 + (V_i^0 I_i^0)^2 + (4V_i^- I_i^0)^2 + \frac{(V_i^0 I_i^-)^2}{4}\right]$$

The remaining two terms contain only nonactive powers

$$(V_i^+ I_{ui})^2 = 9\left[(V_i^+ I_i^-)^2 + 4(V_i^+ I_i^0)^2\right] \quad \text{and} \quad (V_{ui}I_i^+)^2 = 9\left[(V_i^- I_i^+)^2 + \frac{(V_i^0 I_i^+)^2}{4}\right]$$

The rms values of the effective voltage and current for the entire duration τ are

$$V_e = \sqrt{\frac{1}{v}\sum_{i=1}^{v}[(V_i^+)^2 + (V_{ui})^2]} \quad \text{and} \quad I_e = \sqrt{\frac{1}{v}\sum_{i=1}^{v}[(I_i^+)^2 + (I_{ui})^2]}$$

and the effective apparent power squared, calculated for the total observation time τ, has four terms:

$$S_e^2 = (3V_e I_e)^2 = \frac{9}{v^2}\left[\sum_{i=1}^{v}(V_i^+)^2 \sum_{i=1}^{v}(I_i^+)^2 + \sum_{i=1}^{v}(V_i^+)^2 \sum_{i=1}^{v}(I_{ui})^2\right.$$

$$\left. + \sum_{i=1}^{v}(V_{ui})^2 \sum_{i=1}^{v}(I_i^+)^2 + \sum_{i=1}^{v}(V_{ui})^2 \sum_{i=1}^{v}(I_{ui})^2\right]$$

$$= (S^+)^2 + (N_u')^2 + (N_u'')^2 + (S_u')^2 = (S^+)^2 + S_u^2 \tag{7.26}$$

where

$$S_u = \sqrt{(N_u')^2 + (N_u'')^2 + (S_u')^2}$$

The IEEE Std. 1459–2010 [6] calls S_u unbalance power.

1. The first term in (7.26) is the positive-sequence apparent power squared:

$$(S^+)^2 = \frac{9}{v^2}\sum_{i=1}^{v}(V_i^+)^2 \sum_{i=1}^{v}(I_i^+)^2 = \frac{9}{v^2}\left[\left(\sum_{i=1}^{v}V_i^+ I_i^+\right)^2 + \sum_{1\leq n<m\leq v}(V_m^+ I_n^+ - V_n^+ I_m^+)^2\right]$$

$$= (\bar{S}^+)^2 + (D_R^+)^2 = (\bar{P}^+)^2 + (\bar{Q}^+)^2 + (D_R^+)^2 \tag{7.27}$$

where

$$\bar{S}^+ = \frac{1}{v}\sum_{i=1}^{v}S_i^+ ; \qquad S_i^+ = 3V_i^+ I_i^+$$

is the mean positive-sequence apparent power for the duration τ,

$$\bar{P}^+ = \frac{1}{v}\sum_{i=1}^{v}P_i^+ ; \qquad P_i^+ = 3V_i^+ I_i^+ \cos(\theta_i^+)$$

is the mean positive-sequence active power for the duration τ,

$$\bar{Q}^+ = \frac{1}{v}\sum_{i=1}^{v}Q_i^+ ; \qquad Q_i^+ = 3V_i^+ I_i^+ \sin(\theta_i^+)$$

is the mean positive-sequence reactive power for the duration τ, and

$$D_R^+ = \sqrt{\frac{9}{v^2}\left[\sum_{1\leq n<m\leq v}(V_m^+ I_n^+ - V_n^+ I_m^+)^2\right]} \tag{7.28}$$

is the *Positive-Sequence Randomness Power*.

2. The second term is a nonactive power

$$(N_u')^2 = \frac{9}{v^2} \sum_{i=1}^{v} (V_i^+)^2 \sum_{i=1}^{v} (I_{ui})^2 = (\bar{N}_u')^2 + (D_{uIR})^2$$

where

$$\bar{N}_u' = \frac{3}{v} \sum_{i=1}^{v} (V_i^+) I_{ui}$$

is a mean nonactive power due to current unbalance, and

$$D_{uIR} = \sqrt{\frac{9}{v^2} \sum_{1 \le n < m \le v} (V_m^+ I_{un} - V_n^+ I_{um})^2} \tag{7.29}$$

is the *Randomness Power due to Current Unbalance*.
3. The third term

$$(N_u'')^2 = \frac{9}{v^2} \sum_{i=1}^{v} (V_{ui}^+)^2 \sum_{i=1}^{v} (I_i^+)^2 = (\bar{N}_u'')^2 + (D_{uVR})^2$$

is the squared nonactive power due to voltage unbalance, where

$$\bar{N}_u'' = \frac{3}{v} \sum_{i=1}^{v} (V_{ui}) I_i$$

is the mean nonactive power due to voltage unbalance, and

$$D_{uVR} = \sqrt{\frac{9}{v^2} \sum_{1 \le n < m \le v} (V_{um} I_n^+ - V_{un} I_m^+)^2} \tag{7.30}$$

is the *Randomness Power due to Voltage Unbalance*.
4. The last term

$$(S_u')^2 = \frac{9}{v^2} \sum_{i=1}^{v} (V_{ui})^2 \sum_{i=1}^{v} (I_{ui})^2 = (\bar{S}_u')^2 + (D_{uVIR})^2$$

is the unbalance power squared due to voltage unbalance and current unbalance.

$$(\bar{S}_u')^2 = \frac{9}{v^2} \sum_{i=1}^{v} (V_{ui})^2 (I_{ui})^2 = (\bar{P}_u)^2 + (\bar{N}_u''')^2$$

is the mean unbalance apparent power squared, due to unbalance voltage and unbalance current, where

$$\bar{P}_u = \frac{3}{\nu} \sum_{i=1}^{\nu} [V_i^- I_i^- \cos(\theta_i^-) + V_i^0 I_i^0 \cos(\theta_i^0)]$$

is the mean value of the unbalance active power, the sum of negative- and zero-sequence active power

$$\bar{N}_u''' = \sqrt{\frac{9}{\nu^2} \left\{ \sum_{i=1}^{\nu} [(V_i^- I_i^- \sin(\theta_i^-))^2 + (V_i^0 I_i^0 \sin(\theta_i^0))^2 + 4(V_i^- I_i^0)^2 + \frac{(V_i^0 I_i^-)^2}{4}] \right\}}$$

is the mean nonactive power due to voltage and current unbalance. The remaining term

$$D_{uVIR} = \sqrt{\frac{9}{\nu^2} \sum_{1 \le n < m \le \nu} (V_{um} I_{un} - V_{un} I_{um})^2} \tag{7.31}$$

is the *Randomness Power due to Voltage and Current Unbalance*.

Thus, the resolution of the effective apparent power in the case of a sinusoidal, but asymmetrical, system with randomly varying voltages and currents is described by the following expression:

$$S_e^2 = (\bar{S}_e)^2 + (D_R)^2 = (\bar{P}^+)^2 + (\bar{Q}^+)^2 + (\bar{N}_u)^2 + (\bar{P}_u)^2 + (D_R)^2 \tag{7.32}$$

where

$$\bar{N}_u = \sqrt{(\bar{N}_u')^2 + (\bar{N}_u'')^2 + (\bar{N}_u''')^2} \tag{7.33}$$

is the total nonactive power due to the system unbalance and

$$D_R = \sqrt{(D_R^+)^2 + (D_{uIR})^2 + (D_{uVR})^2 + (D_{uVIR})^2} \tag{7.34}$$

is the total *Randomness Power*.

7.5   Three-Phase Systems with Nonsinusoidal and Unbalanced Condition

In this case for every subinterval i the effective voltage and current, V_{ei} and I_{ei}, are separated into two components: the effective fundamental and the effective total harmonic:

$$V_{ei}^2 = V_{e1i}^2 + V_{eHi}^2 \quad \text{and} \quad I_{ei}^2 = I_{e1i}^2 + I_{eHi}^2$$

The effective total harmonics of the current and voltage for the subinterval i are

$$V_{eHi} = \sqrt{\sum_{h\neq 1}^{\infty} V_{ehi}^2} \quad \text{and} \quad I_{eHi} = \sqrt{\sum_{h\neq 1}^{\infty} I_{ehi}^2}$$

where V_{ehi} and I_{ehi} are the rms voltage and current harmonics of h-order. The total effective apparent power squared can be resolved in four terms:

$$S_e^2 = 9(V_e I_e)^2 = \frac{9}{v^2} \sum_{i=1}^{v}(V_e i)^2 \sum_{i=1}^{v}(I_e i)^2$$

$$= \frac{9}{v^2} \sum_{i=1}^{v}(V_{e1i}^2 + V_{eHi}^2) \sum_{i=1}^{v}(I_{e1i}^2 + I_{eHi}^2)$$

$$= (S_{e1})^2 + (D_{eI})^2 + (D_{eV})^2 + (S_{eH})^2 \tag{7.35}$$

1. The first term is the fundamental effective power squared:

$$(S_{e1})^2 = \frac{9}{v^2} \sum_{i=1}^{v}(V_{e1i})^2 \sum_{i=1}^{v}(I_{e1i})^2 = (\bar{P}_1^+)^2 + (\bar{Q}_1^+)^2 + (\bar{N}_{u1})^2 + (D_{e1R})^2 \tag{7.36}$$

Since the fundamental voltages and currents (50 or 60 Hz) are sinusoidal, but not necessarily symmetrical, the observed expressions are identical with (7.34), and the fundamental randomness power is:

$$(D_{e1R})^2 = (D_1^+)^2 + (D_{1uIR})^2 + (D_{1uVR})^2 + (D_{1uVIR})^2$$

2. The second term is the current distortion power squared

$$(D_{eI})^2 = \frac{9}{v^2} \sum_{i+1}^{v}(V_{e1i})^2 \sum_{i+1}^{v}(I_{eHi})^2 = (\bar{D}_{eI})^2 + (D_{eIR})^2 \tag{7.37}$$

where

$$\bar{D}_{eI} = \sqrt{\frac{9}{v^2}(V_{e1i}I_{eHi})^2}$$

is the mean value of the current distortion power for the time τ and

$$D_{eIR} = \sqrt{(D_{eI})^2 - (\bar{D}_{eI})^2} = \sqrt{\frac{9}{v^2} \sum_{1\leq n<m\leq v}^{v}(V_{e1m}I_{eHn} - V_{e1n}I_{eHm})^2}$$

is the *Current Distortion Randomness Power*.

3. The third term is the voltage distortion power squared

$$(D_{eV})^2 = \frac{9}{v^2} \sum_{i+1}^{v} (V_{eHi})^2 \sum_{i+1}^{v} (I_{e1i})^2 = (\bar{D}_{eV})^2 + (D_{eVR})^2 \qquad (7.38)$$

where

$$\bar{D}_{eV} = \sqrt{(D_{eV})^2 - (D_{eVR})^2} = \sqrt{\frac{9}{v^2} (V_{eHi} I_{e1i})^2}$$

is the mean value of the voltage distortion power for the time τ and

$$D_{eVR} = \sqrt{\frac{9}{v^2} \sum_{1 \leq n < m \leq v}^{v} (V_{eHm} I_{e1n} - V_{eHn} I_{e1m})^2}$$

is the *Voltage Distortion Randomness Power*.
4. The fourth term is the harmonic apparent power squared

$$(S_{eH})^2 = \frac{9}{v^2} \sum_{i+1}^{v} (V_{eHi})^2 \sum_{i+1}^{v} (I_{eHi})^2 = (\bar{S}_{eH})^2 + (D_{eHR})^2 \qquad (7.39)$$

where

$$(\bar{S}_{eH})^2 = (\bar{D}_{eH})^2 + (\bar{P}_H)^2 = \frac{9}{v^2} (V_{eHi} I_{eHi})^2$$

is the mean value of the harmonic apparent power for the time τ where

$$D_{eHR} = \sqrt{(S_{eH})^2 - (\bar{S}_{eH})^2} = \sqrt{\frac{9}{v^2} \sum_{1 \leq n < m \leq v}^{v} (V_{eHm} I_{eHn} - V_{eHn} I_{eHm})^2}$$

is the *Harmonic Distortion Randomness Power*
    The harmonic active power $\bar{P}_H = P_H$, is a component of the mean harmonic apparent power $\bar{S}_{eH}$.
    In conclusion, for the most general case the effective apparent power is

$$S_e = \sqrt{(\bar{P}_1^+)^2 + (\bar{Q}_1^+)^2 + (\bar{N}_{u1})^2 + (\bar{P}_u)^2 + (\bar{P}_H)^2 + (\bar{D}_{e1})^2 + (\bar{D}_{eV})^2 + (\bar{D}_{eH})^2 + (D_{eR})^2}$$
$$\qquad (7.40)$$

where

$$N_{u1} = \sqrt{(N'_{u1})^2 + (N''_{u1})^2 + (N'''_{u1})^2}$$

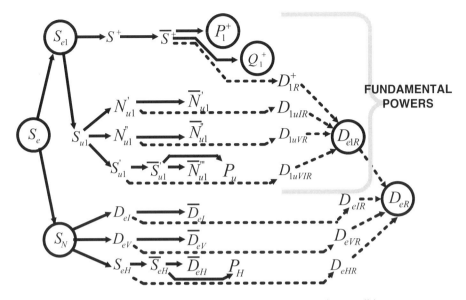

Figure 7.4   Apparent power resolution for time-varying conditions.

and

$$D_{eR} = \sqrt{(D_{e1R})^2 + (D_{eIR})^2 + (D_{eVR})^2 + (D_{eHR})^2} \tag{7.41}$$

is the total *Randomness Power*

The apparent power resolution for the time-varying conditions is shown in Fig. 7.4. The priority powers' symbols were enclosed in circles. One counts three active powers: the significant fundamental active power P_1, and the undesirable harmonic and unbalanced active powers P_H and P_u. There are seven nonactive powers: Q_1, N'_{u1}, N''_{u1}, $\overline{N}'''_{u1}$, D_{eI}, D_{eV}, and D_{eH}. There are also 7 randomness powers that result in the total quantity D_{eR}.

The reader may feel overwhelmed by so many components, but in practice one needs only a few power components to obtain a good picture of the line utilization, harmonic pollution, and the size of filters needed for current distortion correction. Usually S_e, S_N, P_1, Q_1, and D_R, and the total harmonic distortions, are sufficient for a preliminary evaluation of a load or cluster of loads.

In this chapter it is shown that the equivalent values for active and nonactive powers, measured over large time intervals divided into subintervals, are always the mean value of the quantities measured for each subinterval. This feature, however, does not apply to the measurements of apparent power, and it is necessary to include a randomness power in the resolution of apparent power, even when we deal with purely resistive loads.

A simple explanation for these observations can be given by assuming a time-varying resistance supplied with a sinusoidal voltage with constant rms value and a sinusoidal current with randomly varying amplitude. Such a current can be separated into a constant amplitude current and a randomly varying current with zero average value calculated for the total observation time. This random current component causes no net transfer of energy; during some subintervals it helps to transmit energy to the load and during the remaining intervals the load generates an equal amount of energy that is returned to the electric grid. This observation is accurately describing the instantaneous randomness power, it is a shear nonactive power, but a power that causes additional losses in the supplying line conductors and transformer windings.

The random component of current is rich in subsynchronous frequency components. If the measuring instrument, that monitors the load, has the capability to compute the powers for the entire duration τ in a deterministic way, i.e. the subsynchronous components are correctly included in the measurement, then the randomness powers are found to be nil.

It is hard to predict if the randomness power will gain acceptability, nevertheless this new quantity deserves consideration when one deals with arc furnaces, spot welders, and kVA Demand meters.

EXAMPLE: PROBABILISTIC APPROACH

A sinusoidal, 1000V, single-phase voltage, supplies a time-varying load via a feeder with $0.5 + j2 \; \Omega$ equivalent impedance. The load consists of an impedance $R(t) + j50 \; \Omega$. The resistance $R(t)$ varies randomly between the values 50 to 200 Ω according to a normal distribution:

$$p(R, \sigma) = \frac{1}{\sqrt{2\pi}\sigma} \exp\left[-\frac{(R - 100)^2}{2\sigma^2}\right]$$

where the mean value $< R >= 100 \; \Omega$ and the standard deviation is $0 \leq \sigma \leq 60 \; \Omega$.

For a given integer value of R the probability function $p(R, \sigma)$ provides the number of occurrences when the load resistance takes the value R [7]. Thus the rms load voltage and current are as follows

$$V = \sqrt{\int_{50}^{200} \frac{(R^2 + 50^2)10^6}{(R + 0.5)^2 + 52^2} p(R, \sigma) \, dR}$$

$$I = \sqrt{\int_{50}^{200} \frac{10^6}{(R + 0.5)^2 + 52^2} p(R, \sigma) \, dR}$$

In the same manner one finds the active and reactive powers:

$$P = \int_{50}^{200} \frac{10^6}{(R + 0.5)^2 + 52^2} R \, p(R, \sigma) \, dR$$

$$Q = \int_{50}^{200} \frac{10^6}{(R + 0.5)^2 + 52^2} 50 \, p(R, \sigma) \, dR$$

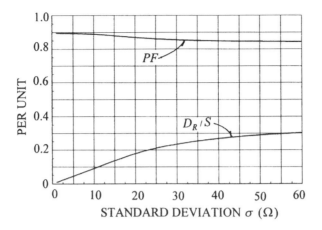

Figure 7.5 Example: Power factor and normalized randomness power versus the standard deviation.

The apparent power is $S = VI$ and the randomness power and the power factor follow:

$$D_R = \sqrt{S^2 - P^2 - Q^2} \quad \text{and} \quad PF = \frac{P}{S}$$

In Fig. 7.5 the power factor and the normalized randomness power as a function of the standard deviation are presented.

These results should be evaluated keeping in mind that the power factor of the mean value impedance $100 + j50$ is $< PF >= 0.894$. This is the value obtained for $\sigma = 0$. As the standard deviation increases the power factor decreases and the randomness power is a significant $D_R = 0.30S$ at $\sigma = 55 \ \Omega$.

7.6 Problems

7.1 A purely resistive load is supplied with a sinusoidal voltage that has a perfectly constant rms value V. A measurement carried over the duration $\tau = 2\Delta\tau$ provided the following information: the rms current $I = I_1$ for $0 \le t \le \Delta\tau$ and $I = I_2$ for $\Delta\tau \le t \le 2\Delta\tau$. Show that the randomness power is $D_R = V|I_1 - I_2|/2$.

7.2 Show that if $V_i = RI_i$ and $R = $ constant, then $\bar{S} = S = P$ and $D_R = 0$.

7.3 Extend the above problem to the general case when the load $\mathbf{Z} = Z\angle\theta$ is constant and the voltage varies in time. Assume v equal time intervals $\Delta\tau$, each characterized by a constant rms voltage V_i with $i = 1, 2, 3, \ldots v$. Compute S, P, Q, D_R, Q_c, and PF.

7.4 An inductive load is monitored over five equal time intervals. The measured quantities are given in Table 7.3. Compute P, $\bar{Q}$, Q_c, S, $\bar{S}$, and D_R.

7.5 A three-phase, Y-connected load with grounded neutral is supplied from a substation via a four-wire system feeder with line resistance $R_S = 0.1 \ \Omega$ per phase and line reactance

Table 7.3　Summary of the Measured Quantities (Problem 7.4)

i	1	2	3	4	5
R_i (Ω)	1.0	0.8	0.5	1.0	2.0
ωL (Ω)	1.0	0.6	0.6	0.8	3.5
V_i (V)	240	235	234	238	242

$\omega L_S = 0.5$ Ω per phase. The neutral path impedance $\mathbf{Z_N} = R_N + jX_N$ varies in time. The load impedances are also time variant. They are labeled as follows:

Phase A: $\mathbf{Z_A} = R_A + jX_A$ Phase B: $\mathbf{Z_B} = R_B + jX_B$ Phase C: $\mathbf{Z_C} = R_C + jX_C$

The line-to-neutral rms voltage phasors at the substation are kept constant and symmetrical. Their rms phasors are:

$\mathbf{V_A} = 440\angle 0^o$ V; $\mathbf{V_B} = 440\angle -120^o$ V; and $\mathbf{V_C} = 440\angle 120^o$ V.

The values of the randomly varying resistances and reactances are summarized in Table 7.4

Table 7.4　Resistance and Reactance Values (Problem 7.5)

i	1	2	3	4	5
R_A (Ω)	2.0	4.0	2.0	3.0	3.0
R_B (Ω)	2.0	5.0	4.0	2.0	4.0
R_C (Ω)	3.0	6.0	4.0	3.0	2.0
R_N (Ω)	4.0	8.0	4.0	6.0	5.0
X_A (Ω)	1.0	0.8	1.0	1.0	1.0
X_B (Ω)	1.0	0.6	0.8	1.5	0.8
X_C (Ω)	1.0	0.4	0.8	1.0	1.0
X_N (Ω)	7.0	6.0	6.6	6.0	6.0

Compute S, P, Q, and D_R.

7.6 The measured voltage and current phasors at the input terminals of a load are given in Table 7.5. V_1 and V_5 are the fundamental and the 5th harmonic voltage. I_1 and I_5 are the fundamental and the 5th harmonic current. Compute S, S_1, P, Q, D_I, D_V, S_H, and D_R.

Table 7.5　Summary of the Measured Quantities (Problem 7.6)

i	1	2	3	4	5
V_1 (pu)	$100\angle 0°$	$105\angle 0°$	$102\angle 0°$	$90\angle 0°$	$97\angle 0°$
V_5 (pu)	$10\angle 40°$	$100\angle 70°$	$100\angle 60°$	$100\angle 45°$	$100\angle 88°$
I_1 (pu)	$100\angle 0°$	$20\angle 0°$	$60\angle 0°$	$120\angle 0°$	$130\angle 0°$
I_5 (pu)	$20\angle 0°$	$35\angle 0°$	$25\angle 0°$	$12\angle 0°$	$4\angle 0°$

7.7 A three-phase load that consists of three equal impedances $\mathbf{Z} = 1 + j1$ pu, $\Delta-$ connected, is supplied with square wave voltages that vary randomly, (see Table 7.6). Compute $\bar{S}_{e1}$, $\bar{D}_{eI}$, $\bar{D}_{eV}$, $\bar{S}_{eH}$, and D_{eR}.

Table 7.6 Summary of the Measured Quantities: Line-to-Line
Voltages. Square Waves Amplitudes, (Problem 7.7)

i	1	2	3	4	5
V_{AB} (pu)	10	12	9	10	11
V_{BC} (pu)	10	10	11	9	9
V_{CA} (pu)	9	9	10	11	12

Note:

$$v\text{-square wave} = \frac{4}{\pi}V\left[\sin(\omega t) + \frac{1}{3}\sin(3\omega t) + \frac{1}{5}\sin(5\omega t) + \cdots\right]$$

7.8 A randomly varying load consists of a resistance that follows a uniform distribution in the range $2 \leq R \leq 198$ Ω with a mean value $< R >= 100\Omega$ and a standard deviation $\sigma = 56.58\Omega$. The supply line has an equivalent impedance $\mathbf{Z} = 0.1 + j0.5$ Ω. Compute S, P, Q, and D_R.

7.9 A three-phase, three-wire line supples a V-connected load with sinusoidal and constant voltage: $\mathbf{V_{AB}} = 220\angle 0°$ V and $\mathbf{V_{BC}} = 220\angle -120°$ V. The two loads, connected line-to-line, have identical impedances, $\mathbf{Z} = R + jX$ Ω, that vary randomly in time following a Gaussian distribution with $< R >= 100$ Ω mean value. The ratio $X/R = 1.2$ remains constant in time. Study the performance of this system in function of the standard deviation $0 \leq \sigma \leq 80$ Ω.

7.7 References

[1] IEC 61000–4–30, Power Quality Measurement Methods.
[2] Edison Electric Institute: "Handbook for Electricity Metering," Washington D. C., 1981.
[3] Emanuel A. E., Orr J. A., Cyganski D., Gulachenski E. M.: "*A Survey of Harmonic Voltages and Currents at Distribution Substations,*" IEEE Transactions on Power Delivery, Vol. 6, No. 4, Oct. 1991, pp. 1883–1890.
[4] Emanuel A. E., Orr J. A., Cyganski D., Gulachenski E. M.: "*A Survey of Harmonic Voltages at the Customer's Bus,*" IEEE Transactions on Power Delivery, Vol. 8, No. 1, Jan. 1993, pp. 411–421.
[5] Emanuel A. E.: "The Randomness Power: An Other New Quantity to be Considered," IEEE Transactions on Power Delivery, Vol. 22, No. 3, July 2007, pp. 1304–1308.
[6] IEEE Standard 1459, "Draft Standard Definitions for the Measurement of Electric Power Quantities Under Sinusoidal, Nonsinusoidal, Balanced, or Unbalanced Conditions," New York, August 2002.
[7] Papoulis A.: "Probability, Random Variables, and Stochastic Processes," McGraw-Hill, Inc. New York, 1991.

8

Appendices

It is not so difficult a task to plant new truths as to root out old errors, for there is this paradox in men: they run after that which is new, but are prejudiced in favor of that which is old.

—Arthur Schopenhauer

8.1 Appendix I: The Electrostatic Field Distribution in a Coaxial Cable

The coaxial cable shown in Fig. 8.1a has an inner conductor with radius a and an outer conductor with an internal radius b. The voltage v_δ drops uniformly along the cable and it is assumed to be equally divided between the outer and the inner conductors. The voltage v_δ supports two electric fields: a radial field caused by the potential difference between the inner and outer conductor and an axial electrostatic field. The total electric field is the result of a superposition represented by Figs. 8.1b and c: In Fig. 8.1b we account for a radial field only, no voltage drop takes place; the same voltage $v_s - v_\delta$ is impressed at both ends of the cable, hence no axial field exists. In Fig. 8.1c only the effect of voltage v_δ is considered and both axial and radial fields are present.

The system being linear, the superposition principle is valid. The computation of the field for Fig. 8.1b is known (1.22), nevertheless finding the field distribution in Fig.8.1c requires a bit of work and to do so we will use the Laplace equation written in cylindrical coordinates [1]. The actual variables are r, the radius, and z, the axial location of the observed point. The potential distribution between the concentrical conductors has the general solution

$$\Phi = Az \ln r + Bz + C \ln r + D \tag{8.1}$$

where A, B, C, and D are constants determined from three boundary conditions (see Fig. 8.1c):

1. For $z = 0$ and $a \leq r \leq b$

$$\Phi = v_\delta/2 = C \ln r + D \tag{8.2}$$

This means that $C = 0$ and $D = v_\delta/2$

Power Definitions and the Physical Mechanism of Power Flow   Alexander Eigeles Emanuel
© 2010 John Wiley & Sons, Ltd

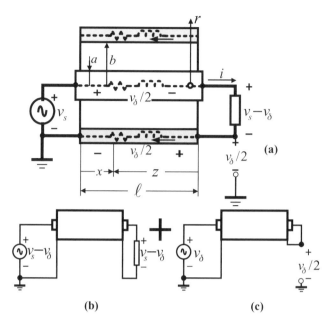

Figure 8.1 Coaxial cable: (a) Actual geometry and voltages. (b) Equivalent system with no-voltage drop. (c) Equivalent sytem accounting for the voltage drop v_δ.

2. For $z = \ell$ and $r = b$

$$\Phi = 0 = A\ell \ln b + B\ell + v_\delta/2 \tag{8.3}$$

3. For $z = \ell$ and $r = a$

$$\Phi = v_\delta = A\ell \ln a + B\ell + v_\delta/2 \tag{8.4}$$

The substitution of (8.3) in (8.4) gives

$$v_\delta = A\ell(\ln a - \ln b) \quad \text{hence} \quad A = \frac{v_\delta}{\ell \ln(a/b)}$$

Substituting the expression of A in (8.3) gives

$$B\ell = -\frac{v_\delta}{2} - \frac{v_\delta}{\ell \ln(a/b)} \ell \ln b \quad \text{hence} \quad B = -\frac{v_\delta}{\ell} \left(\frac{1}{2} + \frac{\ln b}{\ln(a/b)} \right)$$

Substitution of A, B, C, and D in (8.1) gives [1]

$$\Phi = \frac{v_\delta}{\ell} \frac{1}{\ln(a/b)} z \ln r - \frac{v_\delta}{\ell} \left(\frac{1}{2} + \frac{\ln b}{\ln(a/b)} \right) z + \frac{v_\delta}{2}$$

---

[1] This result must be checked:
At $x = \ell$ $\Phi = v_\delta/2$. At $x = 0$ and $r = a$ we find $\Phi = v_\delta$, and at $x = 0$ and $r = b$ $\Phi = 0$.

$$= \frac{v_\delta}{2}\left(1 + \frac{z}{\ell}\frac{\ln(ab/r^2)}{\ln(b/a)}\right) = \frac{v_\delta}{2}\left(1 + \frac{\ell - x}{\ell}\frac{\ln(ab/r^2)}{\ln(b/a)}\right) \qquad (8.5)$$

The potential Φ is a function of x and r and due to axial symmetry is independent of the angle ϕ. This observation leads to the conclusion that the electric field has two components: An axial component

$$\vec{E}'_x = -\frac{\partial \Phi}{\partial x} = \frac{\ln(ab/r^2)}{2\ell \ln(b/a)} v_\delta \vec{1}_x \qquad (8.6)$$

and a radial component

$$\vec{E}'_r = -\frac{\partial \Phi}{\partial r} = -\frac{v_\delta}{2}\frac{\ell - x}{\ell}\frac{(-2/r)}{\ln(b/a)}\vec{1}_r = \frac{1}{\ln(b/a)}\frac{\ell - x}{\ell}v_\delta\frac{1}{r}\vec{1}_r \qquad (8.7)$$

The electric field impressed by the voltages shown in Fig. 8.1b is not a function of x. This is the case of the ideal cylindrical concentric electrodes where the voltage $v_s - v_\delta$ is constant along the cable and the electric field is

$$\vec{E}''_r = \frac{1}{\ln(b/a)}\frac{v_s - v_\delta}{r}\vec{1}_r \qquad (8.8)$$

The total radial field is

$$\vec{E}_r = \vec{E}'_r + \vec{E}''_r = \frac{1}{\ln(b/a)}\left[\frac{\ell - x}{\ell}v_\delta + v_s - v_\delta\right]\frac{\vec{1}_r}{r} = \frac{1}{\ln(b/a)}\frac{v_x}{\ln(b/a)}\frac{\vec{1}_r}{r} \qquad (8.9)$$

where $v_x = v_s - \frac{x}{\ell}v_\delta$ is the cable's voltage at the point x, Fig. 7.1a.

8.2 Appendix II: Poynting Vector due to Displacement Current

Where there is a capacitance C and the capacitance voltage v varies in time there is a displacement current, $i_D = Cdv/dt$, that flows through the dielectric. The coaxial's cable (Fig. 8.1a) capacitance is

$$C = \frac{\varepsilon\ell}{\ln(b/a)}$$

Ignoring the fringing effect at the ends of the cable and assuming $v_\delta \ll v_s$ the displacement current decreases linearly along the cable from $C\,dv_s/dt$ to 0 and produces a magnetic field

$$\vec{H}_D \approx \left(1 - \frac{x}{\ell}\right)\frac{C\,dv_s/dt}{2\pi r}\vec{1}_\phi$$

The voltage v_s supports the electric field

$$\vec{E} \approx \frac{v_s}{\ln(b/a)} \frac{1}{r} \vec{1}_r$$

which interacts with the magnetic field H_D yielding the Poynting vector

$$\vec{\wp}_D = \vec{E} \times \vec{H}_D \approx \frac{1 - x/\ell}{\ln(b/a)} \frac{1}{2\pi r^2} C v_s \frac{dv_s}{dt} \vec{1}_x$$

The power flow carried by the $\wp_D$ is nonactive (has zero mean value). This power flows in and out in the axial direction between the conductors. At $x = 0$, the entry point, the total power is given by the flux carried by $\wp_D$

$$\int_a^b \vec{\wp}_D 2\pi r \, dr \, \vec{1}_x = C v_s \frac{dv_s}{dt} = v_s i_D$$

where i_D is the displacement current.

In situations where the displacement current cannot be ignored the magnetic field in the cable is computed considering both the components, i.e.

$$\vec{H} = \vec{H}' + \vec{H}_D = \frac{1}{2\pi} \left[\frac{i}{r} + \frac{C(1 - x/\ell)}{r} \frac{dv_s}{dt} \right] \vec{1}_\phi$$

8.3 Appendix III: Electric Field Caused by a Time-Varying Magnetic Field

Where there is a magnetic field that varies in time there is always an electric field curling around the magnetic field stream lines. Faraday's law was mathematically expressed by Maxwell and Heaviside as

$$\nabla \times \vec{E}_H = -\mu_0 \frac{\partial \vec{H}}{\partial t} = \frac{-\mu_0}{2\pi r} \frac{di}{dt} \vec{1}_\phi \quad \text{for } a \leq r \leq b$$

The above equation is true for a coaxial cable (Fig. 8.1) where

$$\vec{H} = \frac{i}{2\pi r} \vec{1}_\phi$$

and translated into cylindrical coordinates (r, ϕ, x) has the matrix form

$$\nabla \times \vec{E}_H = \frac{1}{r} \begin{vmatrix} \vec{1}_r & r\vec{1}_\phi & \vec{1}_x \\ \frac{\partial}{\partial r} & \frac{\partial}{\partial \phi} & \frac{\partial}{\partial x} \\ E_{Hr} & rE_{H\phi} & E_{Hx} \end{vmatrix} = \frac{-\mu_0}{2\pi r} \frac{di}{dt} \vec{1}_\phi$$

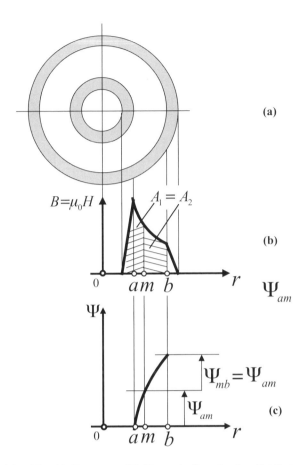

Figure 8.2 Coaxial cable: (a) Geometry. (b) Magnetic flux density distributon. (c) Magnetic flux distribution within the dielectric. The location of the "midway" point m.

Inside a coaxial cable the electric field streamlines, caused by the magnetic field, are parallel to the conductors, the components E_{Hr} and $E_{H\phi}$ are nil, therefore

$$\frac{1}{r}\frac{\partial E_{Hx}}{\partial r} r \, \vec{1}_\phi = \frac{-\mu_0}{2\pi r}\frac{di}{dt} \vec{1}_\phi \qquad (8.10)$$

Integration of (8.10) gives

$$E_{Hx} = \frac{-\mu_0}{2\pi} \ln\frac{r}{K}\frac{di}{dt}$$

The integration constant is found from the condition $E_{Hx} = 0$ at $r = m = K$. The radius m fulfills the condition that the flux Ψ_{am} entering the area $(m - a)\ell$ is equal to the flux Ψ_{mb} through the remaining area, $(b - m)\ell$, Fig. 8.2c. This constrain is written as follows:

$$2\Psi_{am} = \Psi_{ab} \quad \text{or} \quad 2\int_{r=a}^{m} \mu_0 H(r)\ell\,dr = \int_{r=a}^{b} \mu_0 H(r)\ell\,dr$$

thus

$$2\int_{a}^{m} \frac{dr}{r} = \int_{a}^{b} \frac{dr}{r}$$

therefore $2\ln(m/a) = \ln(b/a)$ and $m = \sqrt{ab}$. The axial electric field vector induced by the magnetic field is

$$\vec{E}_{Hx} = \frac{-\mu_0}{2\pi} \ln \frac{r}{\sqrt{ab}} \frac{di}{dt} \vec{1}_x \quad \text{for} \quad a \le r \le b$$

The total axial component of the electric field is

$$\vec{E}_x = \vec{E}_x' + \vec{E}_{Hx} = \left[F(r)v_\delta + G(r)\frac{di}{dt} \right] \vec{1}_x \tag{8.11}$$

where

$$F(r) = \frac{\ln(ab/r^2)}{2\ell\,\ln(b/a)} \quad \text{and} \quad G(r) = \frac{-\mu_0}{2\pi} \ln \frac{r}{\sqrt{ab}}$$

We find that at $r = m = \sqrt{ab}$ the functions $F(m) = G(m) = 0$ yielding $E_x = 0$. The cylindrical surface $2\pi m\ell$ separates two regions; $a < r < m$ and $m < r < b$. The directions of flow of the electric field vector $\vec{E}_H$ in one region is opposed to the direction of flow in the other region. The interaction of the axial electric field with the magnetic field produces a transversal (radial) Poynting vector

$$\vec{\wp}_r = \vec{E} \times \vec{H} = \vec{\wp}_r' + \vec{\wp}_r'' = \left[\frac{\ln(ab/r^2)}{2\ell\,\ln(a/b)} v_\delta \frac{i}{2\pi r} \vec{1}_r - \frac{\mu_0}{2\pi} \ln \frac{r}{\sqrt{ab}} \frac{i}{2\pi r} \frac{di}{dt} \right] \vec{1}_r \tag{8.12}$$

The first component $\vec{\wp}_r'$ carries the active and reactive powers impinged into the conductors, namely the conductors' power loss and the energy stored in the magnetic field located within the conductors. The second term $vec{\wp}_r''$ carries only the reactive power caused by the oscillations of energy stored in and out the dielectric, i.e., the region $a < r < b$.

The total power carried by $\vec{\wp}_r''$ is

$$p_r'' = \frac{-\mu_0}{4\pi^2} \ln \frac{a}{\sqrt{ab}} i\frac{di}{dt} 2\pi a\ell + \frac{\mu_0}{4\pi^2} \ln \frac{b}{\sqrt{ab}} i\frac{di}{dt} 2\pi b\ell$$

$$= \frac{\mu_0\ell}{2\pi} \left(\ln \sqrt{\frac{b}{a}} - \ln \sqrt{\frac{a}{b}} \right) i\frac{di}{dt} = \frac{\mu_0\ell}{2\pi} \ln \left(\frac{b}{a} \right) i\frac{di}{dt} \tag{8.13}$$

The total magnetic flux in the volume $\pi(b^2 - a^2)\ell$ is

$$\Psi = \int_a^b B\ell \, dr = \frac{\mu_0 \ell}{2\pi} i \int_a^b \frac{dr}{r} = \frac{\mu_0 \ell}{2\pi} \ln\left(\frac{b}{a}\right) i = L_H i$$

and this result is in perfect harmony with (8.13) indicating that $p_r^{..} = L_H i \, di/dt$; where L_H is the inductance that is linked with the flux Ψ.

8.4 Appendix IV: The Electromagnetic Wave Along the Three-Phase Line

We will start by assuming a three-phase, perfectly balanced system with the following voltages and currents:

$$
\begin{aligned}
v_a &= \widehat{V} \sin(\omega t) & i_a &= \widehat{I} \sin(\omega t - \theta) \\
v_b &= \widehat{V} \sin(\omega t - 2\pi/3) & i_b &= \widehat{I} \sin(\omega t - 2\pi/3 - \theta) \\
v_c &= \widehat{V} \sin(\omega t + 2\pi/3) & i_c &= \widehat{I} \sin(\omega t + 2\pi/3 - \theta)
\end{aligned}
$$

The instantaneous power is

$$p = v_a i_a + v_b i_b + v_c i_c = p_a + p_q$$

where the active instantaneous power is

$$p_a = \frac{P}{3}[1 - \cos(2\omega t) + 1 - \cos(2\omega t - 4\pi/3) + 1 - \cos(2\omega t + 4\pi/3)] = P$$

and has a constant value equal to the active power $P = 3VI\cos(\theta)$. The instantaneous reactive power

$$p_q = -\frac{Q}{3}[\sin(2\omega t) + \sin(2\omega t - 4\pi/3) + \sin(2\omega t + 4\pi/3)] = 0$$

is nil and seems to have no oscillation in spite of the fact that the reactive power $Q = 3VI\sin(\theta)$ contributes to the line power loss:

$$\Delta P = 3R_s I^2 = \frac{R_s}{3V^2}(P^2 + Q^2)$$

This observation has puzzled engineers for a long time. The correct answer becomes evident if one uses the Poynting vector to understand the distribution of power flow—in time and space—around a polyphase line or cable, Fig. 8.3. The conductors are assumed to be thin, superconductive cylinders with the outer radius a. The conductors A, B, C ... k are assumed to be surrounded by a hypothetical grounded cylinder with a large inner radius $b \gg a$. At a

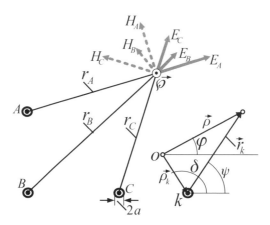

Figure 8.3   Poynting vector $\vec{\wp}$ near a polyphase line.

point $M(\rho, \phi)$, $a \ll \rho \ll b$, the electric field caused by the conductor k (see 2.53) is

$$\vec{E}_k = \frac{\lambda_k}{2\pi\varepsilon_0 r_k} = K_E \frac{v_k}{r_k}\vec{1}_{r_k} ; \quad k = A, \ B, \ C, \dots$$

where λ_k is the line charge, C/m, v_k is the potential of the conductor with respect to ground, ε_0 is the permitivity of air (or of the insulating material that surrounds the conductors). The constant $K_E = 1/\ln(b/a)$ and the radius $\vec{r}_k = \vec{\rho} - \vec{\rho}_k$, see Fig. 8.3.

The magnetic field caused by the conductor k at $M(\rho, \psi)$ is

$$\vec{H}_k = \frac{i_k}{2\pi r_k}\vec{1}_{\psi_k}$$

The Poynting vector at a point $M(\rho, \psi)$ is $\vec{\wp} = \vec{E} \times \vec{H}$ where

$$\vec{E} = k_E \sum_k \frac{v_k}{r_k}\vec{1}_{r_k} ; \quad \vec{H} = \frac{1}{2\pi}\sum_k \frac{i_k}{r_k}\vec{1}_{\psi_k}$$

The total instantaneous power transmitted by this symmetrical polyphase system is

$$p = \int_{\rho=0}^{b}\int_{\psi=0}^{2\pi} \wp\rho \, d\rho \, d\psi = P$$

and is constant in time, but the space and time distribution of $\vec{\wp}$ is more revealing: if we freeze the Poynting vector distribution produced by a three-phase equilateral conductors geometry supplying a unity power factor load at $\omega t = 45°$, Fig. 8.4a, we notice that the Poynting vector is concentrated around the conductors (the Poynting vector is inverse proportionally to the square of the radius measured from the center of the conductor). Also we witness the unidirectional nature of the distribution, typical for $PF = 1$, when the power flows along the conductors, always from the source to the load.

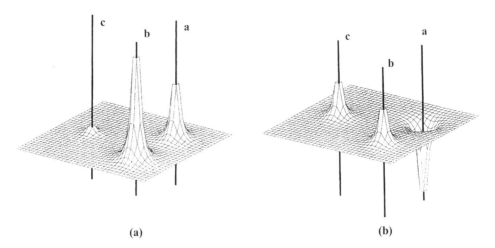

Figure 8.4 Poynting vector envelopes at $\omega t = \pi/4$, near a three-phase line with equilateral conductors placement. The energy flows from bottom of page up toward the load located at the top: (a) Unity power factor load. (b) Pure inductive load, $PF = 0$.

In Fig. 8.4b is shown the Poynting vector distribution at $\omega t = 45°$, for a load with $PF = 0$, (perfectly inductive). This time the flow of power is bidirectional and the total flux of power entering an infinite plane perpendicular on the conductors is nil, i.e.

$$\int_{\rho=0}^{b} \int_{\psi=0}^{2\pi} \wp\rho d\rho d\psi = 0$$

In Fig. 8.5 is shown the Poynting vector distribution at $\omega t = 0°$, $15°$, $30°$, $45°$, and $60°$ ($\omega = 21,600$ deg/s). The distribution pattern repeats every half-cycle and every sixth of a cycle the pattern rotates $120°$.

These three-dimensional plots prove the existence of oscillations of energy associated with both the active and the reactive power. Moreover, considering the fact that when the conductors' impedance is accounted, a transversal Poynting vector component is slightly curving the streamlines toward the conductors' surface and a small amount of Poynting vector flux impinging into conductors is carrying the line power loss as well as the energy stored in the magnetic field distributed within the conductors.

To fathom more in-depth the aspects of the energy flow in three-phase systems we will consider the hypothetical coaxial three-phase cable sketched in Fig. 8.6. The three cylinders (assumed to be very thin and superconductive) have the radii a, b, and c. In the channel CB, $c \leq r \leq b$ the electric and magnetic fields and the Poynting vector instantaneous values are as follows:

$$E_{CB} = \frac{K_{E_{CB}}}{r} v_{CB} \; ; \quad H_{CB} = \frac{1}{2\pi r} i_C \; ; \quad \wp_{CB} = \frac{K_{E_{CB}}}{2\pi r^2} v_{CB} i_C$$

and for the channel BA, $b \leq r \leq a$, we find:

$$E_{BA} = \frac{K_{E_{BA}}}{r} v_{BA} \; ; \quad H_{BA} = \frac{1}{2\pi r}(i_B + i_C) \; ; \quad \wp_{BA} = \frac{K_{E_{BA}}}{2\pi r^2} v_{BA}(i_B + i_C)$$

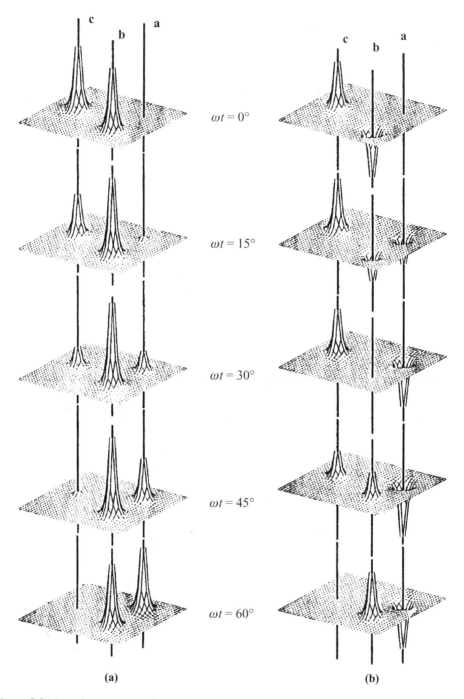

Figure 8.5 Poynting vector envelopes at $\omega t = 0$, $\pi/12$, $\pi/6$, $\pi/4$, $\pi/3$: (a) $PF = 1.0$. (b) $PF = 0$ lagging. Source: Z. Cakareski, A. E. Emanuel, "On the Physical Meaning of Nonactive Powers in Three-phase Systems," IEEE power Engineering Review, July 1999, pp. 46–47.

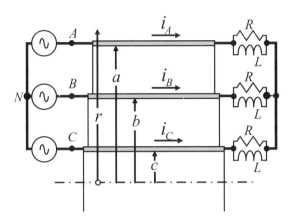

Figure 8.6   Three-phase coaxial cable.

The total instantaneous power transmitted through this cable, computed by means of Poynting vector[2], is:

$$p = \int_{r=c}^{b} \wp_{CB} 2\pi r \, dr + \int_{r=b}^{a} \wp_{BA} 2\pi r \, dr = v_{CB} i_C + v_{BA}(i_B + i_C)$$
$$= (v_C - v_B) i_C + (v_B - v_A)(i_B + i_C)$$
$$= -v_A(i_B + i_c) + v_B(i_B + i_C - i_C) + v_C i_C = v_A i_A + v_B i_B + v_C i_C \qquad (8.14)$$

To observe the actual flow of electromagnetic power through the cable we must substitute the following[3] expressions in (8.14):

$$v_{BA} = -\sqrt{3}\widehat{V}\sin(\omega t + 30°); \quad i_R A = \widehat{I}_R \sin(\omega t); \qquad i_{LA} = \widehat{I}_L \sin(\omega t - 90°)$$
$$v_{CB} = -\sqrt{3}\widehat{V}\sin(\omega t + 90°); \quad i_R B = \widehat{I}_R \sin(\omega t - 120°); \quad i_{LB} = \widehat{I}_L \sin(\omega t + 150°)$$

$$i_{RC} = \widehat{I}_R \sin(\omega t + 120°); \quad i_{LC} = \widehat{I}_L \sin(\omega t + 30°) \qquad (8.15)$$

Thus the active terms are:

$$p_{BA_R} = v_{BA}(i_{RB} + i_{RC}) = \sqrt{3} V I_R[\cos(30°) - \cos(2\omega t + 30°)]$$
$$p_{CB_R} = v_{CB} i_{RC} = \sqrt{3} V I_R[\cos(30°) + \cos(2\omega t + 30°)]$$

---

[2] One should not forget that $i_A + i_B + i_C = 0$, $i_{AR} + i_{BR} + i_{CR} = 0$ and $i_{AL} + i_{BL} + i_{CL} = 0$.

[3] For the sake of clarity the phase shifting was given in degrees instead of radians. Units-wise this is not correct.

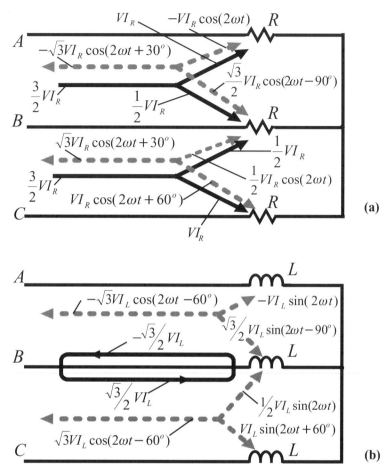

Figure 8.7   Electromagnetic power flow: (a) Toward a balanced unity power factor load. Active power represented by black unidirectional arrows. Intrinsic power by gray bidirectional arrows. (b) Toward an ideal inductive load with $PF = 0$. Reactive power represented by gray bidirectional arrows.

and the reactive

$$p_{BA_L} = v_{BA}(i_{LB} + i_{LC}) = \sqrt{3}VI_L[\cos(120°) - \cos(2\omega t - 60°)]$$
$$p_{CB_L} = v_{CB}i_{LC} = \sqrt{3}VI_L[-\cos(120°) + \cos(2\omega t - 60°)]$$

The correctness of these computations is supported by the results $p_{BA_R} + p_{CB_R} = 3VI_R$ and $p_{BA_L} + p_{CB_L} = 0$, however, the detailed picture is presented in Fig. 8.7, where the actual flow of powers is detailed. The flow of active and intrinsic powers is sketched in Fig. 8.7a and in Fig. 8.7b are presented the nonactive powers. Fig. 8.7b reveals the remarkable fact that the component $(\sqrt{3}/2)VI_L$ "swirls around," in and out of channels AB and BC.

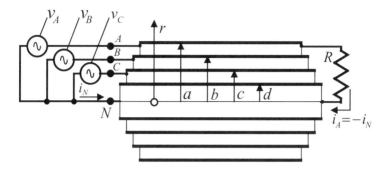

Figure 8.8   Coaxial three-phase cable supplying an unbalanced load.

Next we shall analyze a coaxial three-phase cable that supplies an unbalanced load, Fig. 8.8. The load consists of a resistance R connected between phase A and the neutral N. The supply voltages are symmetrical with the amplitude $\widehat{V}$. The cylinders are assumed to be infinitely thin with the radii a, b, c, and d and made out of superconductive material.

Since the most inner conductor, the neutral, carries current, all the space included in the region $d \leq r \leq a$ carries a magnetic field as well as an electric field. The instantaneous powers supplied by v_B and v_C are nil, nevertheless in the channels $b \leq r \leq c$ and $c \leq r \leq d$ the interaction between the magnetic and electric fields produces a Poynting vector that carries electric power. These instantaneous powers have the following expressions:

$$p_{AB} = \frac{3}{2}VI - \sqrt{3}VI\cos(2\omega t + \pi/6)\,; \quad b \leq r \leq a$$

$$p_{BC} = -\sqrt{3}VI\cos(2\omega t - \pi/2)\,; \qquad c \leq r \leq b$$

$$p_{CN} = -\frac{1}{2}VI - VI\cos(2\omega t + 2\pi/3)\,; \quad d \leq r \leq c$$

An exciting picture is revealed if the first two expressions of the instantaneous powers are rearranged by separating the active and the intrinsic powers as follows:

$$p_{AB} = VI - VI\cos(2\omega t) + \frac{1}{2}VI + VI\cos(2\omega t - 2\pi/3)$$

$$p_{BC} = -VI\cos(2\omega t - 2\pi/3) + VI\cos(2\omega t + 2\pi/3)$$

The power flow, according to the first results, is shown in Fig. 8.9a. The rearranged power flow, after the power separations, is presented in Fig. 8.9b. It is quite unexpected to find that the flux of electromagnetic energy involves loops with no energy transfer. In the channel BC there is a cancellation of Poynting vectors, however, in the channel AB there is an active power component $VI/2$, that flows toward the load and returns via the channel CN. This type of flow of power was pointed to in section 5.4.2, (5.44). This particular example shows the usefulness and the beauty of Poynting vector as a tool that helps to understand the details of the physical mechanisms that govern the flow of electric energy [2].

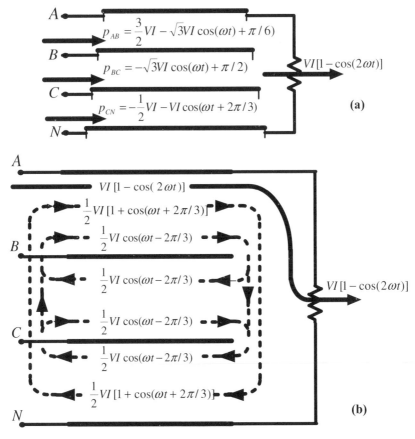

Figure 8.9   Electromagnetic power flow within the cable sketched in Fig. 7.8: (a) First results. (b) After separating the intrinsic and the nonactive powers from the active power.

8.5   Appendix V: Equation (5.99)

The neutral point's potential (Fig. 5.7b) is given in (5.56):

$$v_O = (v_{AN} + v_{BN} + v_{CN})/4 \; ; \qquad v_{N0} = -v_O \qquad (8.16)$$

The collective instantaneous voltage squared (5.62) has the following expression:

$$v_\Sigma^2 = v_{AO}^2 + v_{BO}^2 + v_{CO}^2 + v_{NO}^2 = (v_{AN} - v_O)^2 + (v_{BN} - v_O)^2 + (v_{CN} - v_O)^2 + (v_N - v_O)^2$$

$$= v_{AN}^2 + v_{BN}^2 + v_{CN}^2 + v_N^2 + 4v_O^2 - 2v_O(v_{AN} + v_{BN} + v_{CN} + v_N) \qquad (8.17)$$

Substitution of (8.16) in (8.17) gives

$$v_\Sigma^2 = v_{AN}^2 + v_{BN}^2 + v_{CN}^2 + \frac{4}{16}(v_{AN} + v_{BN} + v_{CN})^2 - \frac{2}{4}(v_{AN} + v_{BN} + v_{CN})^2$$

$$= \frac{1}{4}[3(v_{AN}^2 + v_{BN}^2 + v_{CN}^2) - 2(v_{AN}v_{BN} + v_{BN}v_{CN} + v_{CN}v_{AN})]$$

$$= \frac{1}{4}[v_{AN}^2 + v_{BN}^2 + v_{CN}^2 + (v_{AN} - v_{BN})^2 + (v_{BN} - v_{CN})^2 + (v_{CN} - v_{AN})^2]$$

$$= \frac{1}{4}[v_{AN}^2 + v_{BN}^2 + v_{CN}^2 + v_{AB}^2 + v_{BC}^2 + v_{CA}^2] \tag{8.18}$$

with the rms value

$$V_{\Sigma}^2 = \frac{1}{T}\int_0^T v_{\Sigma}^2 \, dt = \frac{1}{4}[V_A^2 + V_B^2 + V_C^2 + V_{AB}^2 + V_{BC}^2 + V_{CA}^2]$$

8.6 Appendix VI: Maximum Active Power (Three-Phase, Four-Wire System)

The computation of maximum active power by means of Lagrange multipliers can be found in Chapter 5, references [11,12,20,24]. The method described below makes use of conjugate phasors. This approach eliminates tedious computations that involve the phase angles.

Given the line to neutral voltage phasors $\mathbf{V}_A$, $\mathbf{V}_B$, and $\mathbf{V}_C$, and assuming the line power loss for the case $\rho = 1$, the following constraint expressions are involved in this optimization:

$$\mathbf{L} := \frac{\Delta P}{R_s} - (I_A^2 + I_B^2 + I_C^2 + I_N^2) = \frac{\Delta P}{R_s} - (\mathbf{I}_A\mathbf{I}^*_A + \mathbf{I}_B\mathbf{I}^*_B + \mathbf{I}_C\mathbf{I}^*_C + \mathbf{I}_N\mathbf{I}^*_N) = 0 \tag{8.19}$$

and the obvious Kirchhoff's current law

$$\mathbf{M} := \mathbf{I}_A + \mathbf{I}_B + \mathbf{I}_C + \mathbf{I}_N = 0 \tag{8.20}$$

$$\mathbf{N} := \mathbf{I}^*_A + \mathbf{I}^*_B + \mathbf{I}^*_C + \mathbf{I}^*_N = 0 \tag{8.21}$$

The load's active power, that has to be maximized, is

$$P = \frac{1}{2}\Re e\{\mathbf{V}_A\mathbf{I}^*_A + \mathbf{V}_B\mathbf{I}^*_B + \mathbf{V}_C\mathbf{I}^*_C + \mathbf{V}_N\mathbf{I}^*_N + \mathbf{V}^*_A\mathbf{I}_A + \mathbf{V}^*_B\mathbf{I}_B + \mathbf{V}^*_C\mathbf{I}_C + \mathbf{V}^*_N\mathbf{I}_N\} \tag{8.22}$$

The objective function is

$$P + \frac{\lambda}{2}\mathbf{L} - \frac{\mu}{2}\mathbf{M} - \frac{\nu}{2}\mathbf{N} = 0 \tag{8.23}$$

where $\lambda/2$, $\mu/2$ and $\nu/2$ are the respective Lagrange multipliers.

Mathematically P maximization means satisfying the condition

$$\frac{\partial P}{\partial \mathbf{I}_x} - \frac{\lambda}{2}\frac{\partial \mathbf{L}}{\partial \mathbf{I}_x} - \frac{\mu}{2}\frac{\partial \mathbf{M}}{\partial \mathbf{I}_x} - \frac{\nu}{2}\frac{\partial \mathbf{N}}{\partial \mathbf{I}_x} = 0$$

for $\mathbf{I}_x = \mathbf{I}_A$, $\mathbf{I}_B$, $\mathbf{I}_C$, $\mathbf{I}_N$, $\mathbf{I}^*_A$, $\mathbf{I}^*_B$, $\mathbf{I}^*_C$, and $\mathbf{I}^*_N$. This condition yields eight equations for the eight currents. Here are the intermediary computations:

$$\frac{\partial}{\partial \mathbf{I}_A} \rightarrow \quad \mathbf{V}^*_A - \lambda \mathbf{I}^*_A - \mu = 0 \qquad \frac{\partial}{\partial \mathbf{I}^*_A} \rightarrow \quad \mathbf{V}_A - \lambda \mathbf{I}_A - \nu = 0$$

$$\frac{\partial}{\partial \mathbf{I}_B} \rightarrow \quad \mathbf{V}^*_B - \lambda \mathbf{I}^*_B - \mu = 0 \qquad \frac{\partial}{\partial \mathbf{I}^*_B} \rightarrow \quad \mathbf{V}_B - \lambda \mathbf{I}_B - \nu = 0$$

$$\frac{\partial}{\partial \mathbf{I}_A} \rightarrow \quad \mathbf{V}^*_C - \lambda \mathbf{I}^*_C - \mu = 0 \qquad \frac{\partial}{\partial \mathbf{I}^*_C} \rightarrow \quad \mathbf{V}_C - \lambda \mathbf{I}_C - \nu = 0$$

$$\frac{\partial}{\partial \mathbf{I}_N} \rightarrow \quad \mathbf{V}^*_N - \lambda \mathbf{I}^*_N - \mu = 0 \qquad \frac{\partial}{\partial \mathbf{I}^*_N} \rightarrow \quad \mathbf{V}_N - \lambda \mathbf{I}_N - \nu = 0 \qquad (8.24)$$

Using (8.20) and (8.21), the sum of each set of four equations from (8.24) gives

$$\mathbf{V}^*_A + \mathbf{V}^*_B + \mathbf{V}^*_C + \mathbf{V}^*_N - 4\mu = 0 \qquad \mathbf{V}_A + \mathbf{V}_B + \mathbf{V}_C + \mathbf{V}_N - 4\nu = 0$$

or

$$\mathbf{V}_A + \mathbf{V}_B + \mathbf{V}_C + \mathbf{V}_N = 4\mu^* \quad \text{and} \quad \mathbf{V}_A + \mathbf{V}_B + \mathbf{V}_C + \mathbf{V}_N = 4\nu$$

with

$$\nu = \frac{1}{4}(\mathbf{V}_A + \mathbf{V}_B + \mathbf{V}_C + \mathbf{V}_N) \qquad (8.25)$$

and

$$\nu = \mu^* \qquad (8.26)$$

We learned that the multipliers ν and μ are voltage phasors where $\nu = \mathbf{V}_O$, the common mode voltage that defines the reference point O.

The currents that will yield maximum power can now be obtained from (8.24):

$$\mathbf{I}_k = \frac{\mathbf{V}_k - \nu}{\lambda} ; \quad k = A, B, C, N \qquad (8.27)$$

The multiplier λ is a real number, (as a matter of fact it is the resistance $\lambda = R_{\Sigma}$) and is found from (8.19)

$$\frac{\Delta P}{R_s} = (I_A^2 + I_B^2 + I_C^2 + I_N^2)$$

$$= \frac{(\mathbf{V}_A - \nu)(\mathbf{V}_A - \nu)^* + (\mathbf{V}_B - \nu)(\mathbf{V}_B - \nu)^*}{\lambda^2}$$

$$+ \frac{(\mathbf{V}_C - \nu)(\mathbf{V}_C - \nu)^* + (\mathbf{V}_N - \nu)(\mathbf{V}_N - \nu)^*}{\lambda^2}$$

From here we find

$$\lambda = \sqrt{\frac{\sum_k |\mathbf{V}_k - v|^2}{\sum_k I_k^2}} \quad k = A,\ B,\ C,\ N \tag{8.28}$$

Now the maximum power can be calculated:

$$
\begin{aligned}
P_{max} &= \mathfrak{Re}\{\mathbf{V}_A\mathbf{I}^*{}_A + \mathbf{V}_B\mathbf{I}^*{}_B + \mathbf{V}_C\mathbf{I}^*{}_C + \mathbf{V}_N\mathbf{I}^*{}_N\} \\
&= \mathfrak{Re}\{\mathbf{V}_{AO}\mathbf{I}^*{}_A + \mathbf{V}_{BO}\mathbf{I}^*{}_B + \mathbf{V}_{CO}\mathbf{I}^*{}_C + \mathbf{V}_{NO}\mathbf{I}^*{}_N\} \\
&= (\mathbf{V}_A - v)\frac{(\mathbf{V}_A - v)^*}{\lambda} + (\mathbf{V}_B - v)\frac{(\mathbf{V}_B - v)^*}{\lambda} \\
&\quad + (\mathbf{V}_C - v)\frac{(\mathbf{V}_C - v)^*}{\lambda} + (\mathbf{V}_N - v)\frac{(\mathbf{V}_N - v)^*}{\lambda} \\
&= \frac{\sum_k |\mathbf{V}_k - v|^2}{\sqrt{\dfrac{\sum_k |\mathbf{V}_k - v|^2}{\sum_k I_k^2}}} = \sqrt{\sum_k |\mathbf{V}_k - v|^2}\sqrt{\sum_k I_k^2}\ ; \quad k = A,\ B,\ C,\ N \\
&= V_\Sigma I_\Sigma \tag{8.29}
\end{aligned}
$$

with

$$V_\Sigma = \sqrt{\sum_k |\mathbf{V}_k - v|^2} \quad \text{and} \quad I_\Sigma = \sqrt{\sum_k I_k^2}$$

When $\rho \neq 1.0$ the constraint $\mathbf{L}$ (8.19) is modified to

$$\mathbf{L} = \frac{\Delta P}{R_s} - (I_A^2 + I_B^2 + I_C^2 + \rho I_N^2) = \frac{\Delta P}{R_s} - (\mathbf{I}_A\mathbf{I}^*{}_A + \mathbf{I}_B\mathbf{I}^*{}_B + \mathbf{I}_C\mathbf{I}^*{}_C + \rho\mathbf{I}_N\mathbf{I}^*{}_N) = 0 \tag{8.30}$$

and the eight basic equations (8.24) have the terms generated by $\partial/\partial\mathbf{I}_N$ and $\partial/\partial\mathbf{I}^*{}_N$ modified.

$$
\begin{aligned}
\frac{\partial}{\partial\mathbf{I}_A} &\rightarrow & \mathbf{V}^*{}_A - \lambda\mathbf{I}^*{}_A - \mu &= 0 & \frac{\partial}{\partial\mathbf{I}^*{}_A} &\rightarrow & \mathbf{V}_A - \lambda\mathbf{I}_A - v &= 0 \\
\frac{\partial}{\partial\mathbf{I}_B} &\rightarrow & \mathbf{V}^*{}_B - \lambda\mathbf{I}^*{}_B - \mu &= 0 & \frac{\partial}{\partial\mathbf{I}^*{}_B} &\rightarrow & \mathbf{V}_B - \lambda\mathbf{I}_B - v &= 0 \\
\frac{\partial}{\partial\mathbf{I}_C} &\rightarrow & \mathbf{V}^*{}_C - \lambda\mathbf{I}^*{}_C - \mu &= 0 & \frac{\partial}{\partial\mathbf{I}^*{}_C} &\rightarrow & \mathbf{V}_C - \lambda\mathbf{I}_C - v &= 0 \\
\frac{\partial}{\partial\mathbf{I}_N} &\rightarrow & \mathbf{V}^*{}_N - \rho\lambda\mathbf{I}^*{}_N - \mu &= 0 & \frac{\partial}{\partial\mathbf{I}^*{}_N} &\rightarrow & \mathbf{V}_N - \rho\lambda\mathbf{I}_N - v &= 0
\end{aligned}
$$

The sum of the four equations located in the right column gives

$$\mathbf{V}_A + \mathbf{V}_B + \mathbf{V}_C + \frac{\mathbf{V}_N}{\rho} - \lambda(\mathbf{I}_A + \mathbf{I}_B + \mathbf{I}_C + \mathbf{I}_N) = 3v + \frac{v}{\rho}$$

Since $\mathbf{I}_A + \mathbf{I}_B + \mathbf{I}_C + \mathbf{I}_N = 0$ result

$$v = \frac{1}{3 + \dfrac{1}{\rho}}\left(\mathbf{V}_A + \mathbf{V}_B + \mathbf{V}_C + \frac{\mathbf{V}_N}{\rho}\right) \tag{8.31}$$

The currents are

$$\mathbf{I}_k = \frac{\mathbf{V}_k - v}{\lambda}, \quad k = A,\ B,\ C \tag{8.32}$$

and

$$\mathbf{I}_n = \frac{\mathbf{V}_N - v}{\rho\lambda} \tag{8.33}$$

From the line power loss constrain we have

$$R_s(I_A^2 + I_B^2 + I_C^2 + \rho I_N^2) = \frac{\sum_{k=A,B,C}|\mathbf{V}_k - v|^2 + \frac{1}{\rho}|\mathbf{V}_N - v|^2}{\lambda^2}$$

that yields

$$\lambda = \sqrt{\frac{\sum_{k=A,B,C}|\mathbf{V}_k - v|^2 + \frac{1}{\rho}|\mathbf{V}_N - v|^2}{I_A^2 + I_B^2 + I_C^2 + \rho I_N^2}} \tag{8.34}$$

In this case the maximum power is calculated in the same way

$$
\begin{aligned}
P_{max} = S_\Sigma &= \Re e\{\mathbf{V}_{AO}\mathbf{I}^*{}_A + \mathbf{V}_{BO}\mathbf{I}^*{}_B + \mathbf{V}_{CO}\mathbf{I}^*{}_C + \mathbf{V}_{NO}\mathbf{I}^*{}_N\} \\
&= (\mathbf{V}_A - v)\frac{(\mathbf{V}_A - v)^*}{\lambda} + (\mathbf{V}_B - v)\frac{(\mathbf{V}_B - v)^*}{\lambda} \\
&\quad + (\mathbf{V}_C - v)\frac{(\mathbf{V}_C - v)^*}{\lambda} + (\mathbf{V}_N - v)\frac{(\mathbf{V}_N - v)^*}{\rho\lambda} \\
&= \frac{\sum_{k=A,B,C}|\mathbf{V}_k - v|^2 + \frac{1}{\rho}|\mathbf{V}_N - v|^2}{\sqrt{\dfrac{\sum_k|\mathbf{V}_k - v|^2}{\sum_k I_k^2}}} \\
&= \sqrt{\sum_{k=A,B,C}|\mathbf{V}_k - v|^2 + \frac{1}{\rho}|\mathbf{V}_N - v|^2}\sqrt{\sum_{k=A,B,C}I_k^2 + \rho I_N^2} = V_\Sigma I_\Sigma
\end{aligned}
\tag{8.35}
$$

where

$$V_\Sigma = \sqrt{\sum_{k=A,B,C}|\mathbf{V}_k - v|^2 + \frac{1}{\rho}|\mathbf{V}_N - v|^2} = \sqrt{3}V_e'' \tag{8.36}$$

$$I_\Sigma = \sqrt{\sum_{k=A,B,C}I_k^2 + \rho I_N^2} = \sqrt{3}I_e \tag{8.37}$$

8.7   Appendix VII: About the Ratio $\rho = R_s/R_n$

If the skin effect is ignored, a perfectly transposed three-phase line with a phase resistance R_S and neutral path resistance $R_N = \rho R_S$ has the power loss

$$\Delta P_{lines} = R_S(I_A^2 + I_B^2 + I_C^2)$$
$$= 3R_S[(I^+)^2 + (I^-)^2 + (I^0)^2] \qquad (8.38)$$

and the neutral path has the power loss

$$\Delta P_N = R_N I_N^2 = \rho R_S I_N^2 = 9\rho R_S(I^0)^2 \qquad (8.39)$$

From (8.38) and (8.39) we find

$$\rho = \frac{R_N}{R_S} = \frac{\Delta P_N}{I_N^2} \frac{I_A^2 + I_B^2 + I_C^2}{\Delta P_{lines}}$$
$$= \frac{(I^+)^2 + (I^-)^2 + (I^0)^2}{3(I^0)^2} \frac{\Delta P_N}{\Delta P_{lines}} \qquad (8.40)$$

This expression proves that the ratio ρ is a function of line and neutral path losses; however, these losses are not caused solely by the currents of the monitored load, the currents due to other energized loads affect ΔP_{lines} and ΔP_N as well.

A simple example [5], summarized in Fig. 8.10 helps shed light on this nonlinear behavior: A four-wire, three-phase system supplies two separated loads. The current phasors are the following:

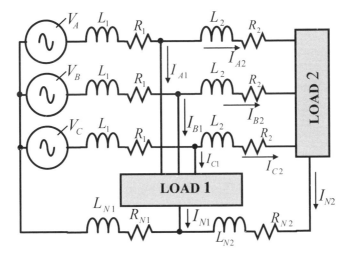

Figure 8.10   Four-wire, three-phase system with two loads.

$$\mathbf{I}_{A1} = I_{A1} \angle \theta_{A1} \; ; \quad \mathbf{I}_{B1} = I_{B1} \angle \theta_{B1} \; ; \quad \mathbf{I}_{C1} = I_{C1} \angle \theta_{C1} \; ; \quad \mathbf{I}_{N1} = I_{N1} \angle \theta_{N1}$$
$$\mathbf{I}_{A2} = I_{A2} \angle \theta_{A2} \; ; \quad \mathbf{I}_{B2} = I_{B2} \angle \theta_{B2} \; ; \quad \mathbf{I}_{C2} = I_{C2} \angle \theta_{C2} \; ; \quad \mathbf{I}_{N2} = I_{N2} \angle \theta_{N2}$$

The total power lost in the lines is

$$\Delta P_{lines} = R_1 \left[|\mathbf{I}_{A1} + \mathbf{I}_{A2}|^2 + |\mathbf{I}_{B1} + \mathbf{I}_{B2}|^2 + |\mathbf{I}_{C1} + \mathbf{I}_{C2}|^2 \right] + R_2 (I_{A2}^2 + I_{B2}^2 + I_{C2}^2)$$
$$= \Delta P_A + \Delta P_B + \Delta P_C \tag{8.41}$$

where

$$\Delta P_A = R_1 I_{A1}^2 + (R_1 + R_2) I_{A2}^2 + 2 R_1 I_{A1} I_{A2} \cos(\theta_{A1} - \theta_{A2}) \tag{8.42}$$

Similar expressions are found for ΔP_B and ΔP_C. The power lost has three terms: the first is due to I_{A1}, the second is due to I_{A2}, and the third is caused by both I_{A1} and I_{A2}. Based on the fact that the power losses are functions of the rms current squared, it is possible to separate the third term into two portions, each one proportional to the I_{A1}^2 and I_{A2}^2, respectively [3,4,5], i.e.

$$2 R_1 I_{A1} I_{A2} \cos(\theta_{A1} - \theta_{A2}) = K_A \left(I_{A1}^2 + I_{A2}^2 \right) R_1$$

where the separation constant is

$$K_A = \frac{2 I_{A1} I_{A2}}{I_{A1}^2 + I_{A2}^2} \cos(\theta_{A1} - \theta_{A2})$$

From (8.42) results that

$$\Delta P_A = \Delta P_{I_{A1}} + \Delta P_{I_{A2}}$$

where

$$\Delta P_{I_{A1}} = (R_1 + K_A R_1) I_{A1}^2 \quad \text{and} \quad \Delta P_{I_{A2}} = (R_1 + R_2 + K_A R_1) I_{A2}^2$$

Similar expressions are obtained for phase B and C. For the neutral path

$$\Delta P_N = \Delta P_{I_{N1}} + \Delta P_{I_{N2}}$$

with

$$\Delta P_{I_{N1}} = (R_{N1} + K_N R_{N1}) I_{N1}^2 \quad \text{and} \quad \Delta P_{I_{N2}} = (R_{N1} + R_{N2} + K_N R_{N1}) I_{N2}^2$$

where

$$K_N = \frac{2 I_{N1} I_{N2}}{I_{N1}^2 + I_{N2}^2} \cos(\theta_{N1} - \theta_{N2})$$

Each load has a theoretical ρ based on (8.40)

$$\rho_1 = \frac{I_{A1}^2 + I_{B1}^2 + I_{C1}^2}{I_{N1}^2} \frac{\Delta P_{N1}}{\Delta P_{A1} + \Delta P_{B1} + \Delta P_{C1}} \tag{8.43}$$

$$\rho_2 = \frac{I_{A2}^2 + I_{B2}^2 + I_{C2}^2}{I_{N2}^2} \frac{\Delta P_{N2}}{\Delta P_{A2} + \Delta P_{B2} + \Delta P_{C2}} \tag{8.44}$$

Practical computations, measurements, or even good estimations of ρ's are not trivial. Random load fluctuations, temperature variations, seasonal changes with humidity and temperature of neutral current paths make the determination of ρ a challenging task for future investigation and research.

In practice $\rho > 1$, and taking $\rho = 1$ does not disadvantage the consumer. In special situations the value of ρ used in S_e and PF measurement can be a matter of agreement between the consumer and utility that supplies electric energy.

A preliminary study [5] indicates that ρ depends mainly on the ratio R_{N2}/R_2. If $R_2/R_1 > 10$, then $\rho \approx R_{N2}/R_2$.

8.8 Appendix VIII: The Use of Varmeters in the Presence of Nonsinusoidal and Asymmetrical Voltages and Currents

Varmeters meant to measure reactive power in systems with sinusoidal waveforms are not the right choice when the current and voltage waveforms are nonsinusoidal [6,7,8,9]. Three-phase dedicated varmeters, which are meant to operate with symmetrical voltages and currents, will err when negative- and zero-sequence voltage and current components are present. Nevertheless, unaware technical personnel may use such meters in the wrong circumstances and reach wrong conclusions. This appendix is meant to shed light on this issue.

There are a multitude of varmeter conceptual designs and each responds differently to nonsinusoidal or asymmetrical conditions. Only the performance of the most common designs is discussed in the following sections.

1. The 90° shift method

We start by assuming clean sinusoidal voltage and current:

$$v(t) = \widehat{V}\cos(\omega t + \alpha); \quad v(t) = \widehat{I}\cos(\omega t + \beta)$$

The integral of the voltage

$$v'(t) = \int \widehat{V}\cos(\omega t + \alpha)\,dt = \frac{\widehat{V}}{\omega}\sin(\omega t + \alpha)$$

is lagging by $90°$ the voltage v, and a wattmeter supplied with the voltage v' and current i will measure

$$P = \frac{1}{kT} \int_0^{kT} v'(t)i(t)\, dt = \frac{\widehat{V}\,\widehat{I}}{kT\omega} \int_0^{kT} \sin(\omega t + \alpha) \cos(\omega t + \beta)\, dt$$

$$= \frac{VIkT}{kT\omega} \sin(\alpha - \beta) = \frac{Q_{int}}{\omega}$$

thus

$$Q_{int} = \frac{\omega}{kT} \int_0^{kT} i(t) \left[\int v\, dt \right] dt = Q = VI \sin(\theta), \quad \theta = \alpha - \beta \tag{8.45}$$

Using the same approach one may build a varmeter using the derivative of voltage that will produce a signal

$$v''(t) = \frac{dv(t)}{dt} = -\omega \widehat{V} \sin(\omega t + \alpha)$$

that is leading by $90°$ the voltage v. In this case the varmeter will be based on the expression:

$$Q_{dif} = \frac{-1}{\omega kT} \int_0^{kT} i(t) \left[\frac{dv(t)}{dt} \right] dt = Q = VI \sin(\theta) \tag{8.46}$$

Next we consider nonsinusoidal voltage and current waveforms:

$$v(t) = V_0 + \widehat{V}_1 \cos(\omega t + \alpha_1) + \cdots + \widehat{V}_h \cos(\omega t + \alpha_h) \tag{8.47}$$
$$i(t) = V_0 + \widehat{I}_1 \cos(\omega t + \beta_1) + \cdots + \widehat{I}_h \cos(\omega t + \beta_h) \tag{8.48}$$

Substitution of (8.47) and (8.48) in (8.45) gives

$$Q_{int} = \omega \left[V_0 I_0 \frac{kT}{2} + \frac{V_1 I_1}{\omega} \sin(\alpha_1 - \beta_1) + \cdots + \frac{V_h I_h}{h\omega} \sin(\alpha_h - \beta_h) \right]$$

$$= \mathbf{Q}_1 + \Delta Q_{int} \tag{8.49}$$

where using the notations $P_0 = V_0 I_0$ and $\omega T = 2\pi$, results

$$\Delta Q_{int} = k\pi P_0 + \sum_{h \neq 1} \frac{Q_h}{h}$$

and $Q_h = V_h I_h \sin(\alpha_h - \beta_h); h = 1, 2, 3, \ldots$
Substitution of (8.47) and (8.48) in (8.46) gives

$$Q_{dif} = V_1 I_1 \sin(\alpha_1 - \beta_1) + \cdots + h V_h I_h \sin(\alpha_h - \beta_h)$$

$$= \mathbf{Q}_1 + \Delta Q_{dif} \tag{8.50}$$

where

$$\Delta_{dif} = \sum_{h \neq 1} h Q_h$$

When dc traces are present the term $k\pi P_0$ may increase ΔQ_{int} significantly if the integration time kT is large. When the current transducer is a current transformer or a Rogowski coil the dc component I_0 is not transferred and the term P_0 is harmless. If Hall effect transducers are used then dc traces may increase ΔQ_{int}, ΔQ_{qcv} and ΔQ_{qci} (see (8.49), (8.52) and (8.53)).

2. The quarter-cycle delay method

This method follows the same concept as the $90°$ shift but the implementation is by digital means. The governing equation for voltage shift is

$$Q_{qcv} = \frac{1}{kT} \int i(t) v(t - T/4) \, dt \qquad (8.51)$$

The shifted voltage signal has the expression

$$v(t - T/4) = V_0 + \widehat{V}_1 \cos\left(\omega t - \frac{\omega T}{4} + \alpha_1\right) + \cdots + \widehat{V}_h \cos\left(h\omega t - \frac{h\omega T}{4} + \alpha_h\right)$$

$$= V_0 + \widehat{V}_1 \cos\left(\omega t - \frac{\pi}{2} + \alpha_1\right) + \widehat{V}_2 \cos(2\omega t - \pi + \alpha_2)$$

$$+ \widehat{V}_3 \cos\left(3\omega t - \frac{3\pi}{2} + \alpha_3\right) + \widehat{V}_4 \cos(4\omega t + \alpha_4) + \cdots$$

and the voltage-current product

$$v(t - T/4) i(t) = V_0 I_0 + V_1 I_1 \cos\left(\alpha_1 - \beta_1 - \frac{\pi}{2}\right) + V_2 I_2 \cos(\alpha_2 - \beta_2 - \pi)$$

$$+ V_3 I_3 \cos\left(\alpha_3 - \beta_3 - \frac{3\pi}{2}\right) + V_4 I_4 \cos(\alpha_4 - \beta_4) + \cdots$$

leads to expression:

$$Q_{qcv} = \mathbf{Q}_1 + \Delta Q_{qcv} \qquad (8.52)$$

where the difference

$$\Delta Q_{qcv} = P_o - P_2 - Q_3 + P_4 + Q_5 - P_6 - \cdots$$

A quarter-cycle current delay will enable the design of a varmeter based on the expression

$$Q_{qci} = \frac{-1}{kT} \int_0^{kT} v(t) i(t - T/4) \, dt$$

with

$$i(t - T/4) = I_0 + \widehat{I_1} \cos\left(\omega t - \frac{\pi}{2} + \beta_1\right) + \widehat{I_2} \cos(2\omega t - \pi + \beta_2)$$
$$+ \widehat{I_3} \cos\left(3\omega t - \frac{3\pi}{2} + \beta_3\right) + \widehat{I_4} \cos(4\omega t + \beta_4) + \cdots$$

resulting in

$$Q_{qci} = \mathbf{Q}_1 + \Delta Q_{qci} \tag{8.53}$$

with

$$\Delta Q_{qci} = -P_0 + P_2 - Q_3 - P_4 + Q_5 + P_6 - \cdots$$

3. Budeanu's method

The harmonic voltage and current phasors are obtained by means of FFT thus enabling the computation of fundamental and harmonic reactive powers.

$$Q_B = \sum_h V_h I_h \sin(\theta_h) = \mathbf{Q}_1 + \Delta Q_B \tag{8.54}$$

where

$$\Delta Q_B = \sum_{h \neq 1} Q_h$$

4. Vector reactive power

Using the vector apparent power definition one may define a reactive power

$$Q_V = \sqrt{S_V^2 - P^2} \tag{8.55}$$

When the apparent power squared is correctly defined, as in (4.52):

$$S_V^2 = V^2 I^2 = P_1^2 + \mathbf{Q}_1^2 + D_I^2 + D_V^2 + P_H^2 + D_H^2$$

The substitution of the total active power squared

$$P^2 = P_1^2 + P_H^2 + 2P_1 P_H$$

in (8.55) gives

$$Q_V = \sqrt{\mathbf{Q}_1^2 + 2P_1 P_H + D_I^2 + D_V^2 + D_H^2} \tag{8.56}$$

This result shows that both quantities Q_1 and Q_V are lacking a physical foundation and practical use.

5. Numerical examples

We will assume a load supplied with the following voltage and current harmonic phasors (in per unit):

$$\mathbf{V_1} = 1.0\angle 0°; \quad \mathbf{V_3} = 0.03\angle 0°; \quad \mathbf{V_5} = 0.03\angle 0°; \quad \mathbf{V_7} = 0.03\angle 0°; \quad \mathbf{V_9} = 0.03\angle 0°$$

and

$$\mathbf{I_1} = 1\angle 90°; \quad \mathbf{I_3} = \frac{1\angle 90°}{3^a}; \quad \mathbf{I_5} = \frac{1\angle 90°}{5^a}; \quad \mathbf{I_7} = \frac{1\angle 90°}{7^a}; \quad \mathbf{I_9} = \frac{1\angle 90°}{9^a}$$

The chosen spectra crudely simulate the fact that many Thévenin impedances, that consist of a simple series $R - L$ circuit, are characterized by $h\omega L \gg R$, hence the harmonic active power is considerably smaller than the harmonic reactive power, i.e. $P_h = RI_h^2 \ll Q_h = h\omega L I_h^2$. In a first approximation it is possible to neglect the active currents. Moreover, the even order active power components, $P_2, P_4, P_6 \dots$ are usually one or two orders of magnitude smaller than the reactive powers. In this example the percent total harmonic distortion of the voltage is $\%THD_V = 6.0\%$. The percent total harmonic distortion of the current is

$$\%THD_I = \sqrt{\left(\frac{1}{3}\right)^{2a} + \left(\frac{1}{5}\right)^{2a} + \left(\frac{1}{7}\right)^{2a} + \left(\frac{1}{9}\right)^{2a}}$$

The exponent a helps adjust the total harmonic distortion of the current. The normalized differences $\Delta Q/Q_1$ are summarized in Table 8.1.

From the obtained results it can be concluded that, with certain indulgence, the varmeters included in this brief survey may be considered to measure the fundamental reactive power. This is especially true for the quarter-cycle delay methods.

In case of near resonance conditions one harmonic becomes dominant. For example let us assume the fifth harmonic large enough, such that all other harmonics can be neglected; $\mathbf{V_5} = 0.05\angle 0°$ per unit and $\mathbf{I_5} = 0.60\angle 270°$ per unit, thus $Q_5 = 0.05 \times 0.60 = 0.03$.

Table 8.1 Calculated percent differences ΔQ, for different types of varmeters.

a	0.6	0.8	1.0	1.2	1.4	1.6	1.8	2.0
$\%THD_I$	76.2	56.8	42.9	32.7	25.2	19.6	15.3	12.0
$\%\Delta Q_B$	4.4	3.2	2.4	1.7	1.3	1.0	0.7	0.6
$\%\Delta Q_{int}$	1.0	0.7	0.6	0.4	0.3	0.2	0.2	0.1
$\%\Delta Q_{dif}$	24.1	17.0	12.0	8.5	6.1	4.4	3.2	2.4
$\%\Delta Q_{qcv}$	0.04	−0.3	−0.5	−0.6	−0.6	−0.6	−0.6	−0.6
$\%\Delta Q_{qci}$	0.04	−0.3	−0.6	−0.6	−0.6	−0.6	−0.6	−0.6

Assuming $Q_1 = 1.0$, the different varmeters will yield the following normalized differences:

$$100\frac{\Delta Q_B}{Q_1} = 3.0\%, \quad 100\frac{\Delta Q_{int}}{Q_1} = 1.0\%, \quad 100\frac{\Delta Q_{dif}}{Q_1} = 15.0\%$$

$$100\frac{\Delta Q_{qcv}}{Q_1} = 100\frac{\Delta Q_{qci}}{Q_1} = 3.0\%$$

In this situation even the quarter-cycle delay varmeters yield readings that are not accurate enough.

From a practical point of view, considering the technology available today, such varmeters are obsolete. The fundamental reactive power can be readily obtained using digital signal processors which perform Fourier analysis (FFT methods). Implementation of such algorithms enables the measurement of voltage and current harmonic and interharmonic phasors, up to a high harmonic order; moreover, also the remaining nonactive powers D_I, D_V and D_H, components that are incorporated in the basic expressions (4.52)–(4.58),

$$S = \sqrt{S_1^2 + S_N^2} = \sqrt{P_1^2 + Q_1^2 + S_N^2} = \sqrt{P_1^2 + \mathbf{Q}_1^2 + D_I^2 + D_V^2 + S_H^2}$$

can be conveniently measured.

The old methods, described in the above subsections, fail to give any information on S_N or its components. This condition can be easily detected in the above numerical example, where for $a = 0.60$, the rms voltage and current (in per unit) are

$$V = 1.002, \quad I = 1.257$$

The apparent power is $S = VI = 1.26$. The active powers $P_1 = P_H = 0$ and the nonactive powers are as follows: $Q_1 = 1.0$, $D_I = 0.58$, $D_V = 0.0036$ and $S_H = 0.021$. The described methods lead to reactive power measurements (depending on the method) in the range $0.995 < Q < 1.241$ while $S = 1.26$ and $P = 0$ remain unchanged. The Pythagorean rule of the right-angle triangle, S, P, Q, is infringed. Instruments based on this method give insufficient and erroneous information. Only for conditions where the voltage and the current waveforms are mildly distorted ($THD_I < 0.1$ and $THD_V < 0.03$) does the expression $S^2 = P^2 + Q^2$ holds true and some of the old methods remain valid. Methods based on the derivative of voltage or current must be totally avoided.

6. Three-phase varmeters

Some three-phase reactive power meters measurements use varmeters, others are based on the time averaging the product between voltage and current when one of the $90°$ voltage shifted phasors is readily available in the existing system. Traditionally varmeter designers took advantage of the fact that a symmetrical three-phase voltage source has the line-to-line voltage phasor $\mathbf{V_{AB}}$ perpendicular to the line-to-neutral phasor $\mathbf{V_{CN}}$, $\mathbf{V_{BC}} \perp \mathbf{V_{AN}}$ and $\mathbf{V_{CA}} \perp \mathbf{V_{BN}}$. This fact enabled the use of wattmeters with the voltage terminals connected to a line-to-line

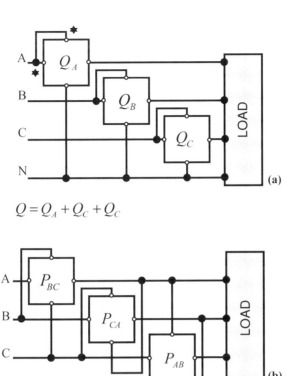

$$Q = Q_A + Q_C + Q_C$$

$$Q = (P_{BC} + P_{CA} + P_{AB})/\sqrt{3}$$

Figure 8.11 Traditional (old) methods for the reactive power measurement: (a) Using three varmeters. (b)) Using three wattmeters (recommended for systems with symmetrical voltages).

voltage 90° out of phase with the line-to-neutral voltage of the line where the current terminals are connected.

There are quite a few designs favored by the industry. In Fig. 8.11 are described the basic concepts of two commonly used methods. The three varmeters method is presented in Fig. 8.11a. The reading is

$$Q = Q_A + Q_B + Q_C = Q_1 + \Delta Q$$

where

$$Q_1 = Q_{A1} + Q_{B1} + Q_{C1}$$

and

$$\Delta Q = \Delta Q_A + \Delta Q_B + \Delta Q_C$$

The discussions and the conclusions presented in sections 1 to 5 applies unmodified to this method, however, the reactive power $Q_1 \neq Q_1^+$, and the value of the fundamental positive-sequence reactive power Q_1^+, makes an essential data for the evaluation of power flow conditions.

In Fig. 8.11b is sketched the conceptual design of the three wattmeters method, which is meant to be used for systems with perfectly symmetrical voltages. The accuracy of the three wattmeters method is affected by both voltage unbalance and waveform distortion. The errors' origin can be detected if one separates the line currents into fundamental positive- and negative-sequence, and the harmonic currents, i.e.

$$i_A = i_A^+ + i_A^- + i_{AH}; \quad i_B = i_B^+ + i_B^- + i_{BH}; \quad i_C = i_C^+ + i_C^- + i_{CH}$$

The same grouping is done for the line-to-line voltages

$$v_{BC} = v_{BC}^+ + v_{BC}^- + v_{BCH}, \quad v_{CA} = v_{CA}^+ + v_{CA}^- + v_{CAH}, \quad v_{AB} = v_{AB}^+ + v_{AB}^- + v_{ABH}$$

where

$$v_{AB}^+ = \widehat{V}^+ \cos(\omega t + 30°); \qquad\qquad i_A^+ = \widehat{I}^+ \cos(\omega t - \theta^+)$$
$$v_{BC}^+ = \widehat{V}^+ \cos(\omega t - 90°); \qquad\qquad i_B^+ = \widehat{I}^+ \cos(\omega t - 120° - \theta^+)$$
$$v_{CA}^+ = \widehat{V}^+ \cos(\omega t + 150°); \qquad\qquad i_C^+ = \widehat{I}^+ \cos(\omega t + 120° - \theta^+)$$
$$v_{AB}^- = \widehat{V}^- \cos(\omega t - 30° + \gamma); \qquad\quad i_A^- = \widehat{I}^- \cos(\omega t - \theta^- + \gamma)$$
$$v_{BC}^- = \widehat{V}^- \cos(\omega t + 90° + \gamma); \qquad\quad i_B^- = \widehat{I}^- \cos(\omega t + 120° - \theta^- + \gamma)$$
$$v_{CA}^- = \widehat{V}^- \cos(\omega t - 150° + \gamma); \qquad\quad i_C^- = \widehat{I}^- \cos(\omega t - 120° - \theta^- + \gamma)$$
$$v_{BCH} = \sum_{h\neq1} \widehat{V}_{BCh} \cos(h\omega t + \alpha_{BCh}); \quad i_{AH} = \sum_{h\neq1} \widehat{I}_{Ah} \cos(h\omega t + \beta_{Ah})$$
$$v_{CAH} = \sum_{h\neq1} \widehat{V}_{CAh} \cos(h\omega t + \alpha_{CAh}); \quad i_{BH} = \sum_{h\neq1} \widehat{I}_{Bh} \cos(h\omega t + \beta_{Bh})$$
$$v_{ABH} = \sum_{h\neq1} \widehat{V}_{ABh} \cos(h\omega t + \alpha_{ABh}); \quad i_{CH} = \sum_{h\neq1} \widehat{I}_{Ch} \cos(h\omega t + \beta_{Ch})$$

The angle γ equals the rotation of the negative-sequence phasor diagram with respect to the positive-sequence diagram. The positive-sequence phase A, line-to-neutral voltage, is the chosen reference.

This instrument measures the average power

$$P = \frac{1}{kT} \int_0^{kT} (p_{BC} + p_{CA} + p_{AB})\, dt \qquad (8.57)$$

where

$$p_{BC} = v_{BC}^+ i_A^+ + v_{BC}^- i_A^- + v_{BC}^+ i_A^- + v_{BC}^- i_A^+ + v_{BCH} i_{AH}$$
$$p_{CA} = v_{CA}^+ i_B^+ + v_{CA}^- i_B^- + v_{CA}^+ i_B^- + v_{CA}^- i_B^+ + v_{CAH} i_{BH}$$
$$p_{AB} = v_{AB}^+ i_C^+ + v_{AB}^- i_C^- + v_{AB}^+ i_C^- + v_{AB}^- i_C^+ + v_{ABH} i_{CH}$$

The cross product terms due to positive-sequence voltage and negative-sequence current, or vice versa, yield nil powers [10].

Computations of the three terms of (8.57) give

$$\frac{1}{kT} \int_0^{kT} p_{BC} \, dt = \sqrt{3}(Q_{BC}^+ + Q_{BC}^-) + P_{BCH}$$

$$\frac{1}{kT} \int_0^{kT} p_{CA} \, dt = \sqrt{3}(Q_{CA}^+ + Q_{CA}^-) + P_{CAH}$$

$$\frac{1}{kT} \int_0^{kT} p_{AB} \, dt = \sqrt{3}(Q_{AB}^+ + Q_{AB}^-) + P_{ABH}$$

where

$$Q_{BC}^+ = Q_{CA}^+ = Q_{AB}^+ = V^+ I^+ \sin(\theta^+)$$
$$Q_{BC}^- = Q_{CA}^- = Q_{AB}^- = V^- I^- \sin(\theta^-)$$

are the positive- and the negative-sequence fundamental reactive powers, respectively, and

$$P_{BCH} = \sum_{h \neq 1} V_{BCh} I_{Ah} \cos(\alpha_{BCh} - \beta_{Ah})$$

$$P_{CAH} = \sum_{h \neq 1} V_{CAh} I_{Bh} \cos(\alpha_{CAh} - \beta_{Bh})$$

$$P_{ABH} = \sum_{h \neq 1} V_{ABh} I_{Ch} \cos(\alpha_{ABh} - \beta_{Ch})$$

are harmonic powers that contribute to the corruption of the measurement.

The final expression for the the three wattmeters method under unbalanced and nonsinusoidal conditions is

$$Q = \frac{1}{\sqrt{3}}(P_{BC} + P_{CA} + P_{AB}) = Q^+ + Q^- + P_H$$

where

$$Q^+ = 3V^+ I^+ \sin(\theta^+); \quad Q^- = 3V^- I^- \sin(\theta^-)$$

and

$$P_H = P_{BCH} + P_{CAH} + P_{ABH}$$

may be a main source of errors in spite of positive and negative values that lead to the partial cancellations of components.

8.9   References

[1] Kraus J. D.: "Electromagnetics," McGraw-Hill, 1984.

[2] Emanuel A. E.: "About the Rejection of Poynting Vector in Power Systems Analysis," *Electrical Power Quality and Utilization Journal*, Vol. XIII, No.1, 2007, pp.41–47.

[3] Emanuel A. E.: "Harmonic Cost Allocation: A Difficult Task," Proceedings of IEEE PES Summer Meeting, July 1999, pp.1115–25.

[4] Davis E. J., Emanuel A. E., Pileggi D. J.: "Harmonic Pollution Metering: Theoretical Considerations," *IEEE transactions on Power Delivery*, Vol.15, No.1, Jan.2000, pp.14–18.

[5] Pajic, S., Emanuel, A. E.: "Effect of Neutral Path Power Losses on the Apparent Power Definitions: A Preliminary Study," *IEEE transactions on Power Delivery*, Vol.24, No.2, April.2009, pp.517–23.

[6] Filipski P. S., Baghzouz Y., Cox M. D.: "Discussion of Power Definitions Contained in the IEEE Dictionary," *IEEE transactions on Power Delivery*, Vol.9, No.3, July 1994, pp.1237–44.

[7] Filipski P. S., Labaj P. W.: "Evaluation of Reactive Power Meters in the Presence of High Harmonic Distortion," *IEEE PES Winter Meeting*, 1992, New York, NY. Paper 92 WM 192–5 PWRD.

[8] Cox M. D., Williams T. B.: "Induction Varhour and Solid-State Varhour Meter Performance on Nonlinear Loads," *IEEE transactions on Power Delivery*, Vol.5, No.4, Nov. 1990, pp.1678–84.

[9] Arseneau R., Filipski P. S.: "Application of a Three-Phase Nonsinusoidal Calibration System for Testing Energy and Demand Meters Under Simulated Field Conditions," *IEEE transactions on Power Delivery*, Vol.PWD–3, July 1988, pp.874–879.

[10] Filipski P. S., Baghzouz Y., Cox M. D.: "Discussion of Power Definitions Contained in the IEEE Dictionary," *IEEE transactions on Power Delivery*, Vol.9, No.3, July 1994, pp.1237–44.

Index